A-LEVEL
STUDENT GUIDE

PEARSON EDEXCEL

Geography

Physical geography

Cameron Dunn
Michael Witherick

HODDER
EDUCATION
AN HACHETTE UK COMPANY

Although every effort has been made to ensure that website addresses are correct at time of going to press, Hodder Education cannot be held responsible for the content of any website mentioned in this book. It is sometimes possible to find a relocated web page by typing in the address of the home page for a website in the URL window of your browser.

Hachette UK's policy is to use papers that are natural, renewable and recyclable products and made from wood grown in well-managed forests and other controlled sources. The logging and manufacturing processes are expected to conform to the environmental regulations of the country of origin.

Orders: please contact Hachette UK Distribution, Hely Hutchinson Centre, Milton Road, Didcot, Oxfordshire, OX11 7HH. Telephone: +44 (0)1235 827827. Email education@hachette.co.uk Lines are open from 9 a.m. to 5 p.m., Monday to Friday. You can also order through our website: www.hoddereducation.co.uk

ISBN: 978 1 3983 2816 7

© Cameron Dunn and Michael Witherick 2021

First published in 2021 by
Hodder Education,
An Hachette UK Company
Carmelite House
50 Victoria Embankment
London EC4Y 0DZ

www.hoddereducation.co.uk

Impression number 10 9 8 7 6 5 4 3

Year 2025 2024 2023

Cover photo: Kalyakan – stock.adobe.com

Typeset in India

Printed and bound by CPI Group (UK) Ltd, Croydon, CR0 4YY

A catalogue record for this title is available from the British Library.

Contents

Content Guidance

Questions and Answers

■ Getting the most from this book

Exam tips

Advice on key points in the text to help you learn and recall content, avoid pitfalls, and polish your exam technique in order to boost your grade.

Knowledge check

Rapid-fire questions throughout the Content Guidance section to check your understanding.

Knowledge check answers

1 Turn to the back of the book for the Knowledge check answers.

Summaries

■ Each core topic is rounded off by a bullet-list summary for quick-check reference of what you need to know.

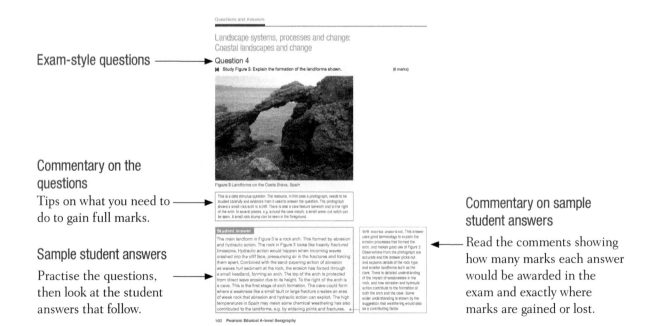

Exam-style questions

Commentary on the questions
Tips on what you need to do to gain full marks.

Sample student answers
Practise the questions, then look at the student answers that follow.

Commentary on sample student answers
Read the comments showing how many marks each answer would be awarded in the exam and exactly where marks are gained or lost.

About this book

Much of the knowledge and understanding needed for A-level geography builds on what you have learned for GCSE geography, but with an added focus on key geographical concepts, depth of knowledge and understanding of content. This guide offers advice for the effective revision of **physical geography**, which all students need to complete.

A-level Paper 1 tests your knowledge and application of physical geography, focusing on **Tectonic processes and hazards**, either **Glaciated landscapes and change** or **Coastal landscapes and change**, the **Water Cycle and water insecurity** and the **Carbon cycle and energy security**. The whole exam lasts 2 hours and 15 minutes and makes up 30% of the A-level qualification. More information on the external exam papers is given in the Questions and Answers section at the back of this book.

To be successful in this unit you have to understand:
■ the key ideas of the content
■ the nature of the assessment material — by reviewing and practising sample structured questions
■ how to achieve a high level of performance within the exams.

This guide has two sections:

Content Guidance — this summarises some of the key information that you need to know to be able to answer the exam questions with a high degree of accuracy and depth. In particular, the meaning of key terms is made clear and details of case study material are provided to help meet the spatial context requirement within the specification.

Synoptic links have also been highlighted to help you make connections between concepts and themes and the three synoptic themes outlined in the specification. These are:

1 Players (P)
2 Actions and Attitudes (A)
3 Futures and Uncertainties (F)

In many cases these synoptic links also link to specialised concepts such as adaptation, mitigation, resilience and globalisation which are relevant within several different topics.

Questions and Answers — this includes sample questions similar in style to those you might expect in the exam. There are sample student responses to these questions as well as detailed analysis, which will give further guidance about what exam markers are looking for to award top marks.

The best way to use this book is to read through the relevant topic area first before practising the questions. Only refer to the answers and examiner comments after you have attempted the questions.

Content Guidance

This section outlines the following areas of the AS geography and A-level geography specifications:

- Tectonic processes and hazards
- Landscape systems, processes and change
  - Glaciated landscapes and change
  - Coastal landscapes and change
- The water cycle and water insecurity
- The carbon cycle and energy security

Read through the topic area before attempting a question from the Questions and Answers section.

■ Tectonic processes and hazards

Why are some locations more at risk from tectonic hazards?

- Tectonic hazards (earthquakes, volcanic eruptions and tsunami) occur in specific locations, related to tectonic plate boundaries and other tectonic settings.
- Their distribution is uneven, with some areas at high risk and other locations at no risk.
- Tectonic events can generate multiple hazards when they occur.

The global distribution of tectonic hazards

All tectonic hazards are caused by the Earth's internal heat engine. Radioactive decay of isotopes such as uranium-238 and thorium-232 in the Earth's core and mantle generate huge amounts of heat which flow towards the Earth's surface. This heat flow generates convection currents in the plastic mantle. The interior of the Earth is therefore dynamic rather than static. Most tectonic hazards occur at or near tectonic plate boundaries. These represent the locations of ascending (divergent plate boundaries) and descending (convergent plate boundaries) arms of mantle convection cells (Figure 1).

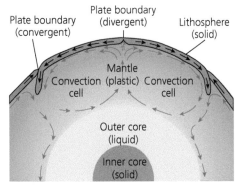

Figure 1 Earth's internal structure and mantle convection

> **Exam tip**
>
> All tectonic hazards are physical events, with natural causes. Only use the word 'disaster' when referring to the impact of these events on people.

> **Exam tip**
>
> The terms 'plate boundary' and 'plate margin' are used interchangeably. They both mean the narrow, linear zone where two tectonic plates meet.

The mantle is a solid, but because of the very high temperatures present it is locally deformable (plastic) and capable of very slow 'flow'.

Figure 2 shows the distribution of earthquakes, volcanoes and tectonic plate boundaries. Most earthquakes occur at, or close to, these boundaries. This is also true of volcanic eruptions. Some plate boundary earthquakes cause a secondary tectonic hazard, tsunami.

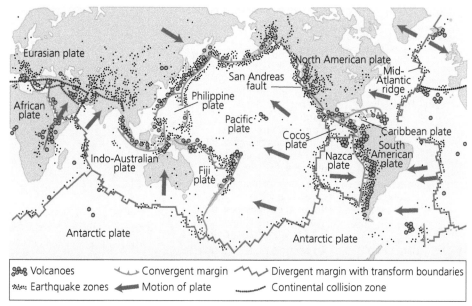

Figure 2 The global distribution of plate boundaries, earthquakes and volcanoes

Not all tectonic plate boundaries are the same and this has an important impact on the type and magnitude of tectonic hazards. Plate boundary type depends on two factors:

1 Motion: whether plates are moving apart (divergent), colliding (convergent) or sliding past each other (conservative or transform).

2 Plate type: whether the tectonic plates are oceanic or continental. Oceanic plates make up the ocean floor and are high-density, basaltic rock but only 7–10 km thick. Continental plates make up Earth's landmasses and are much thicker at 25–70 km but made of less dense granitic rock.

This combination of motion and plate type causes the different plate boundary types that are summarised in Table 1 on p. 8.

Although the vast majority of earthquakes and eruptions occur at plate boundaries, a small number do not. Some volcanic eruptions are described as 'intra-plate'. This means they are distant from a plate boundary at locations called mid-plate hotspots (such as Hawaii and the Galapagos Islands). At these locations:

- isolated plumes of convecting heat, called **mantle plumes**, rise towards the surface, generating basaltic volcanoes that tend to erupt continually
- a mantle plume is stationary, but the tectonic plate above moves slowly over it
- over millennia, this produces a chain of volcanic islands, with extinct ones most distant from the plume location.

Mantle plumes are concentrated areas of heat convection. At plate boundaries they are sheet-like, whereas at hot spots they are column-like.

Table 1 Different plate boundary settings

Divergent (constructive)	Oceanic–Oceanic	Mid-Atlantic ridge at Iceland	Rising convection currents bring magma to the surface, resulting in small, basaltic eruptions, creating new oceanic plate. Minor, shallow earthquakes
	Continent–Continent	African Rift Valley/Red Sea	Caused by a geologically recent mantle plume splitting a continental plate to create a new ocean basin. Basaltic volcanoes and minor earthquakes
Convergent (destructive)	Continent–Continent	Himalayas	The collision of two continental landmasses creating a mountain belt as the landmasses crumple. Infrequent major earthquakes distributed over a wide area
	Oceanic–Oceanic	Aleutian Islands, Alaska	One oceanic plate is subducted beneath another, generating frequent earthquakes and a curving (arc) chain of volcanic islands with violent eruptions
	Oceanic–Continent	Andean Mountains	An oceanic plate is subducted under a continental plate, creating a volcanic mountain range, frequent large earthquakes and violent eruptions
Conservative	Continent–Continent	California, San Andreas fault zone	Plates slide past each other, along zones known as transform faults. Frequent, shallow earthquakes but no volcanic activity

Exam tip

Make sure you know which types of plate (oceanic/continental) are involved in each of your examples of plate boundaries.

Knowledge check 1

State two differences between oceanic and continental plates.

Earthquakes can occur in mid-plate settings, usually associated with major ancient fault lines being reactivated by tectonic stresses or areas of crustal weakness and thinning. For instance, the New Madrid Seismic Zone on the Mississippi River generates earthquakes up to magnitude 7.5 but is thousands of miles from the nearest plate boundary.

Theories of plate motion

Earth's tectonic plates move at a speed of 2–5 cm per year. There are seven very large major plates (African, Pacific), smaller minor plates (Nazca, Philippine Sea) and dozens of small microplates. All fit together into a constantly moving jigsaw of rigid lithosphere. Each plate is about 100 km thick. Its lower part consists of upper mantle material while its upper part is either oceanic or continental crust.

The theory of plate tectonics has developed because of a number of key discoveries:
- Alfred Wegener's Continental Drift hypothesis in 1912 that postulated that now-separate continents had once been joined.
- The ideas of Arthur Holmes in the 1930s that Earth's internal radioactive heat was the driving force causing mantle convection that could move tectonic plates.

- The discovery in 1960 of the asthenosphere, a weak, deformable layer beneath the rigid lithosphere on which the lithosphere moves.
- The discovery in the 1960s of magnetic stripes in the oceanic crust of the sea bed; these are palaeomagnetic (ancient magnetism) signals from past reversals of Earth's magnetic field and prove that new oceanic crust is created by the process of sea-floor spreading at mid-ocean ridges.
- The recognition of transform faults by Tuzo Wilson in 1965.

It remains a theory because scientists have not yet directly observed the interior of the Earth.

Knowledge check 2

Why is plate tectonics still a theory rather than proven fact?

Constructive margins

Mantle convection forces plates apart at constructive plate margins. Tensional forces open cracks and faults between the two plates. These create pathways for magma to move towards the surface and erupt, creating new oceanic plate. Eruptions are small and effusive in character, as the erupted basalt lava has a low gas content and low viscosity. Earthquakes are shallow, less than 60 km deep, and have low magnitudes, usually under 5.0.

Destructive margins and subduction zones

Locations where one plate is subducted beneath another illustrate the forces that drive plate tectonics (Figure 3).

- Mantle convection pulls oceanic plates apart, creating the fracture zones at constructive margins, and convection also pulls plates towards subduction zones.
- Constructive margins have elevated altitudes because of the rising heat beneath them, which creates a 'slope' down which oceanic plates slide (gravitational sliding or 'ridge push').
- Cold, dense oceanic plate is subducted beneath less dense continental plate; the density of the oceanic plate pulls itself into the mantle (slab pull).

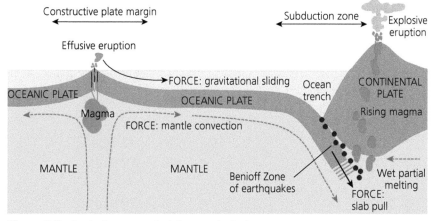

Figure 3 Tectonic forces and plate margin features

Earthquakes at **subduction** zones occur at a range of focal depths from 10 km to 400 km, following the line of the subducting plate. This is called a Benioff Zone and it can yield very large earthquakes up to magnitude 9.0. The descending plate begins to melt

Subduction is the process of one plate sinking beneath another at a convergent (destructive) plate boundary.

at depth by a process called wet partial melting. This generates magma with a high gas and silica content, which erupts with explosive force.

Collision zones

The Himalaya mountains is a location where two continental plates are in collision (the Indo-Australian and Eurasian plates). The collision began about 52 million years ago. As both continental plates have the same low density, subduction is not possible. Instead, the plates have 'crumpled', creating enormous tectonic uplift in the form of the Himalayan and Tibetan Plateau. Magma is being generated at depth, but it cools and solidifies beneath the surface so eruptions are rare. Collision zones are cut by huge thrust faults that generate shallow, high-magnitude earthquakes such as in Kashmir in 2005 and Nepal in 2015.

Transform zones

Conservative plate boundaries consist of transform faults. These faults 'join up' sections of constructive plate boundary as they traverse the Earth's surface in a zig-zag pattern. In some locations, long transform faults act like a boundary in their own right, most famously in California where a fault zone — including the San Andreas fault — creates an area of frequent earthquake activity. Earthquakes along conservative boundaries often have shallow focal depths, meaning high-magnitude earthquakes can be very destructive. Volcanic activity is absent.

The causes of tectonic hazards

Earthquakes

Earthquakes are a sudden release of stored energy. As tectonic plates attempt to move past each other along fault lines, they inevitably 'stick'. This allows strain to build up over time and the plates are placed under increasing stress. Earthquakes are generated because of sudden stress release — so-called 'stick-slip' behaviour. A pulse of energy radiates out in all directions from the earthquake **focus** (point of origin, sometimes called the hypocentre) (Figure 4). In some cases the earthquake motion displaces the surface, so a fault scarp is formed.

> **Exam tip**
>
> Sometimes it can be quicker and easier to sketch and annotate a diagram of a plate boundary in the exam.

An earthquake originates at the **focus**. The epicentre is the point on the Earth's surface directly above the focus.

> **Knowledge check 3**
>
> What is an earthquake?

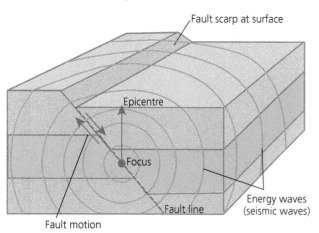

Figure 4 The anatomy of an earthquake

Earthquakes generate three types of seismic wave:

■ P-waves, or primary waves, are the fastest. They arrive first and cause the least damage.
■ S-waves, or secondary waves, arrive next and shake the ground violently, causing damage.
■ L-waves, or Love waves, arrive last as they travel only across the surface. However, they have a large amplitude and cause significant damage, including fracturing the ground surface.

Earthquakes cause crustal fracturing within the Earth, but also buckle and fracture the ground surface. Some very large earthquakes, such as the one that generated the 2004 Indian Ocean tsunami, rupture a fault line for up to 1000 km. Think of this as 'unzipping' a fault, with energy pulses being generated along the entire fault length. Such earthquakes can cause ground shaking that lasts up to 5 minutes, as well as dozens of powerful aftershocks.

Earthquakes frequently generate large landslides as secondary hazards. This is especially true in areas of geologically young (and therefore unstable) mountains such as the Himalayas. Landslides accounted for up to 30% of deaths in the 2008 Sichuan and 2005 Kashmir earthquakes.

Liquefaction is a particular hazard in areas where the ground consists of loose sediment such as silt, sand or gravel that is also waterlogged — often found in areas close to the sea or lakes. Intense earthquake shaking compacts the loose sediment together, forcing water between the sediment out and upward. This undermines foundations and causes buildings to sink, tilt and often collapse.

Volcanoes

Major volcanic eruptions frequently have more than one hazard associated with them. In some cases these are secondary hazards, which are an indirect consequence of the eruption (lahar, jökulhlaup). This is especially the case with the violent eruptions associated with volcanoes at destructive plate margins, as shown in Table 2 on p. 12.

In most cases, only large composite volcanoes found at destructive plate margins represent a significant tectonic hazard. These eruptions often have lava flows, pyroclastic flows, lahars and extensive ash and tephra fall that can affect areas up to 30 km from the volcanic vent.

Pyroclasts (meaning 'fire broken') are any rock fragments ejected from a volcano, including ash, tephra and volcanic bombs.

Tsunami

There have been a number of deadly tsunami in recent years (Table 3, p. 12) such that this hazard is more well known than it once was. Tsunami can be generated by landslides and even eruptions of volcanic islands. Most are generated by sub-marine earthquakes at subduction zones.

Most tsunami are generated when a sub-marine earthquake displaces the sea bed vertically (either up or down) as a result of movement along a fault line at a subduction zone. The violent motion displaces a large volume of water in the ocean water column, which then moves outward in all directions from the point of displacement. The water moves as a vast 'bulge' in open water, rather than as a distinct wave.

Table 2 Volcanic hazards

Volcanic hazard	Explanation	Volcano types
Lava flow	Extensive areas of solidified lava, which can extend several kilometres from volcanic vents if the lava is basaltic and low viscosity. It can flow at up to 40 kmh	✚ ⬤
Pyroclastic flow	Very large, dense clouds of hot ash and gas at temperatures of up to 600°C. They can flow down the flanks of volcanoes and devastate large areas	✚
Ash fall	Ash particles, and larger tephra particles, can blanket huge areas in ash, killing vegetation, collapsing buildings and poisoning water courses	▪ ✚
Gas eruption	The eruption of carbon dioxide and sulphur dioxide, which can poison people and animals in extreme cases	✚ ⬤
Lahar	Volcanic mudflows, which occur when rainfall mobilises volcanic ash. They travel at high speed down river systems and cause major destruction	✚
Jökulhlaup	Devastating floods caused when volcanoes erupt beneath glaciers and ice caps, creating huge volumes of meltwater. They are common in Iceland	▪

✚ Subduction zone volcano (composite type)

⬤ Hot-spot volcano (shield type)

▪ Constructive plate margin volcano (cinder cone, fissure eruption)

Table 3 Deadly tsunami since 2004

Location	Earthquake magnitude	Wave height (m)	Deaths
2004 Indian Ocean	9.2	24	230,000
2006 Java, Indonesia	7.7	2–6	800
2009 Samoa	8.1	14	190
2011 Japan	9.0	9.3	16,000
2018 Indonesia	Volcanic eruption of Anak Krakatau	2–5	425

Tsunami characteristics are very different from those of wind-generated ocean waves:

- wave heights are typically less than 1 m
- wavelengths are usually more than 100 km
- speeds are 500–950 km/h.

In the open ocean tsunami waves are barely noticeable. As the waves approach shore they slow dramatically, wavelength decreases but wave height increases. Tsunami usually hit coastlines as a series of waves (a 'wave-train'), the effect of which is more like a flood than a breaking wave. Sub-marine earthquakes that occur close to shorelines can generate intense ground-shaking damage, followed by damage from the subsequent tsunami.

Knowledge check 4

Which type of plate boundary produces the most hazardous volcanoes?

Exam tip

Tsunami have nothing to do with tides, despite being commonly called 'tidal waves'.

Knowledge check 5

What has to happen to the sea bed during an earthquake for a tsunami to be generated?

Why do some tectonic hazards develop into disasters?

- There is a relationship between hazard, vulnerability and resilience that sometimes leads to tectonic disasters.
- The impacts of disasters vary hugely between locations.
- Tectonic disasters and their impacts can be understood using hazard profiles.
- Level of development has a major influence on the nature of disasters.
- Governance is a key concept in understanding the impacts of disasters.

> **Exam tip**
>
> Use key words from the specification in your answers. Although related, the word 'governance' has a different meaning to 'management'.

Hazard versus disaster

Tectonic hazards are natural events that have the potential to harm people and their property. A disaster is the realisation of a hazard, i.e. harm has occurred. By definition disasters have to involve people and they occur at the intersection of people and hazards, as shown by the Degg disaster model (Figure 5).

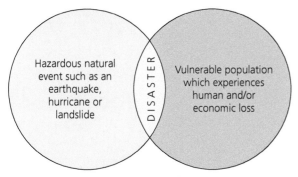

Figure 5 The Degg disaster model

A **threshold** level is often used to determine whether the impact of an event is large enough to be considered a disaster, such as:

- 10 or more deaths
- 100 or more people affected
- US$1 million in economic losses.

We can understand the relationship between hazards and disasters using the hazard risk equation. The risk of disaster rises if hazard magnitude rises, i.e. a VEI 6 eruption compared with a VEI 3. If vulnerability rises (poverty, lack of preparedness, lack of awareness of potential hazards), so does risk:

$$\text{risk} = \frac{\text{hazard} \times \text{vulnerability}}{\text{capacity to cope}}$$

A **threshold** is the magnitude above which a disaster occurs. This threshold level could be different in a developed versus a developing country because of different levels of resilience.

Some communities have a high capacity to cope and high **resilience**. This means they can reduce the chances of disasters occurring because:

- they have emergency evacuation, rescue and relief systems in place
- they react by helping each other to reduce numbers affected
- hazard-resistant design or land-use planning have reduced the numbers at risk.

For these communities the threshold for a disaster will be higher than for ones with low coping capacity.

Some disasters are truly catastrophic in terms of their impact. A good example is the 2010 Port-au-Prince earthquake in Haiti. Its magnitude of 7.0 was relatively low, but the death toll has been estimated at 160,000. The Pressure and Release model (PAR model) can be used to help understand the enormous death toll (Figure 6).

Progression of vulnerability				Natural hazards
→ **Root causes**	→ **Dynamic pressures**	→ **Unsafe conditions**	Disaster (risk = hazard × vulnerability)	← Earthquake
Low access to resources	Lack of education, training and investment	Poor construction standards		Eruption
Limited influence in decision making		Unsafe infrastructure		Tsunami
	Rapid population change and urbanisation	Poverty		
Poor governance and a weak economic system		Lack of social safety net		

Figure 6 The Pressure and Release model
Source: after Blaikie (1994)

The PAR model suggests that the socioeconomic context of a hazard is important. In poor, badly governed (root causes) places with rapid change (dynamic pressures) and low coping capacity (unsafe conditions), disasters are likely. Sadly, in 2010 Haiti fitted many of these criteria (Table 4).

Table 4 The PAR model applied to the 2010 Haiti earthquake

Per capita GDP (PPP) US$1200	70% of jobs are in the informal sector	25% of people live in extreme poverty
50% of the population is under 20 years old	Port-au-Prince, a city of 3.5 million, has no sewer system	80% of Port-au-Prince's housing is unplanned, informal slums

The impacts of tectonic hazards

The impacts of tectonic hazards are broadly of three types:

1 Social: deaths, injury and wider health impacts, including psychological ones.
2 Economic: the loss of property, businesses, infrastructure and opportunity.
3 Environmental: damage or destruction of physical systems, especially ecosystems.

Resilience is the ability of a community to cope with a hazard; some communities are better prepared than others so a hazard is less likely to become a disaster.

Exam tip

You can draw models, such as the Degg disaster model, in your answers to exam questions.

Knowledge check 6

What is the name given to the outcome between a vulnerable population and a natural hazard?

In the last 30 years, different tectonic hazards have had contrasting impacts in terms of scale (Table 5).

Table 5 Contrasting tectonic hazard impacts

Volcanic eruptions	Earthquakes	Tsunami
Small and declining impacts, especially death tolls	Large impacts, as major earthquakes are common and widespread	Very large impacts from a small number of events

Comparing impacts between countries is difficult because both the physical nature of the event and the socioeconomic profiles of affected places are different. Table 6 on p. 16 compares the impact of recent disasters in developed, developing and emerging countries. Some general observations are:

■ economic costs in developed and emerging economies are, in some cases, enormous
■ deaths in developed countries are low, except for the 2011 Japanese tsunami (a rare mega-disaster)
■ volcanic eruption impacts are small compared with those of earthquakes and tsunami.

Measuring magnitude and intensity

The 'size' of a tectonic event, called its magnitude, has a relationship with its impact. Broadly, larger magnitude events have a bigger impact — but the relationship is not a simple one because of the vulnerability and capacity to cope parts of the hazard risk equation.

Earthquake magnitude is measured using the moment magnitude scale (MMS). This is an updated version of the well-known Richter magnitude scale. MMS measures the energy released during an earthquake. This is related to the amount of slip (movement) on a fault plane and the area of movement on the fault plane. MMS uses a logarithmic scale, meaning that a magnitude 6 earthquake has ten times more ground shaking than a magnitude 5.

The Mercalli scale measures earthquake intensity on a scale of I–XII. This older scale measures what people actually feel during an earthquake, i.e. the intensity of the shaking effects, not the energy released. It cannot be easily used to compare earthquakes as shaking experienced depends on building type and quality, ground conditions and other factors.

The relationship between magnitude and death toll is a weak one because:

■ some earthquakes cause serious secondary impacts, such as landslides and tsunami
■ earthquakes hitting urban areas have greater impacts than those in rural areas
■ level of development, and level of preparedness, affect death tolls
■ isolated, hard-to-reach places could have a higher death toll because rescue and relief take longer.

The magnitude of a volcanic eruption is measured using the Volcanic Explosivity Index (VEI). VEI ranges from 0 to 8 and is a composite index combining eruption height, volume of material (ash, gas, tephra) erupted and duration of eruption (Table 7 on p. 17).

A **mega-disaster** is a disaster with unusually high impacts. Today that means millions of people affected and billions of dollars in damage over a wide area, i.e. an entire region or even more than one country.

Knowledge check 7

What type of scales are the Richter and moment magnitude scales?

Exam tip

Learn some key facts and figures about your examples and case studies as these add weight to your answers.

Exam tip

Make sure you understand the difference between magnitude and intensity.

Table 6 The impact of disasters compared

	Developed countries	Emerging countries	Developing countries
Volcanic eruption	2010 Eyjafjallajökull (Iceland) Constructive margin, mid-ocean ridge Basaltic magma; stratovolcano VEI = 4 ■ No injuries, no deaths ■ Major disruption to European and transatlantic air travel affecting 10 million passengers and costing US$1.7 billion in economic losses ■ Ice melt on the volcano caused some flash flooding	2010 Merapi (Indonesia) Destructive margin subduction zone Andesitic magma; composite cone volcano VEI = 4 ■ 353 deaths, about 500 injured ■ 350,000 people successfully evacuated before the eruption ■ US$0.6 billion in losses ■ Loss of rice harvest due to ash fall and some destruction of forests due to pyroclastic flows	2002 Nyiragongo (DRC) Constructive margin, continental rift zone Basaltic magma, stratovolcano VEI = 1 ■ 147 deaths, 120,000 made homeless ■ 15% of the city of Goma destroyed by lava flows ■ Major international aid response launched ■ US$1.2 billion in economic losses
Earthquake	2010 Canterbury (New Zealand) Magnitude 7.1 Focal depth = 10 km Subduction zone ■ 100 injuries, no deaths ■ Widespread building damage due to liquefaction ■ A magnitude 6.3 aftershock in 2011 in Christchurch killed 185 ■ Total costs (including aftershocks) estimated at US$40 billion	2008 Sichuan (China) Magnitude 8.0 Focal depth = 19 km Continent–Continent collision zone ■ 69,000 deaths, 370,000 injured ■ At least 5 million homeless ■ More than US$140 billion in economic losses ■ About one-third of deaths due to landslides	2015 Gorkha (Nepal) Magnitude 7.9 Focal depth = 8.2 km Continent–Continent collision zone ■ 9000 deaths, 22,000 injured ■ Economic losses about US$5 billion ■ Killer avalanches triggered on Mt Everest ■ Many rural villages totally destroyed
Tsunami	2011 Tohoku (Japan) Magnitude 9.0 9.3 m tsunami height Megathrust subduction zone ■ 16,000 deaths and 6000 injuries ■ US$300 billion in economic losses ■ 46,000 buildings destroyed and 145,000 damaged ■ Huge infrastructure damage to ports, water and electricity supply	2018 Sulawesi (Indonesia) Magnitude 7.5 7 m tsunami height Strike-slip fault zone ■ 4,340 deaths ■ 205,000 made homeless ■ Economic costs of US$1–2 billion ■ Violent ground-shaking, tsunami and liquefaction all contributed to the high impacts	2004 Indian Ocean Magnitude 9.2 24 m tsunami height Megathrust subduction zone ■ 230,000 deaths and 125,000 injured ■ 1.7 million people displaced across 15 countries ■ US$15 billion in economic losses

Liquefaction occurs in waterlogged, loose sediment; earthquake shaking 'liquefies' the ground, causing buildings to tilt, sink and collapse.

Table 7 The Volcanic Explosivity Index scale

VEI	0	1	2	3	4	5	6	7	8
Eruption height	<100 m	100 m–1 km	1–5 km	3–15 km	>10 km	>10 km	>20 km	>20 km	>20 km
Eruption volume	<10,000 m³	>10,000 m³	>0.001 km³	>0.01 km³	>0.1 km³	>1 km³	>10 km³	>100 km³	>1000 km³
	Effusive			Explosive				Colossal	

VEI eruptions from 0 to 3 are associated with shield volcanoes and basaltic eruptions at constructive plate margins and mid-plate hotspots. VEI eruptions from 4 to 7 occur at destructive plate margins, erupting high-viscosity, high-gas, high-silica andesitic magma. No modern human has experienced a VEI 8 supervolcano. These are rare caldera eruptions such as Yellowstone and Toba.

A **supervolcano** is one whose impacts would be felt globally, because of a worldwide cooling of the Earth's climate, perhaps for up to 5 years.

Hazard profiles

Tectonic events can be compared using hazard profiles. These allow a better understanding of the nature of hazards and therefore risks associated with each. Figure 7 shows hazard profiles for four tectonic hazards. Hazards with the following characteristics present the highest risk:

- high magnitude, low frequency events — these are the least 'expected' as, by definition, they are unlikely to have occurred in living memory
- rapid onset events with low spatial predictability — they could occur in numerous places, and happen without warning
- regional areal extent — affecting large numbers of people in a wide range of locations.

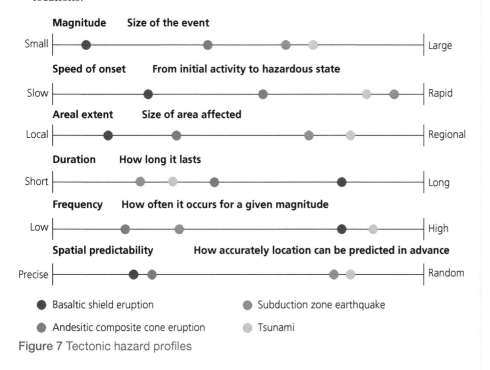

Figure 7 Tectonic hazard profiles

Arguably, major earthquakes at subduction zones and collision zones are the most dangerous tectonic hazards. They can have magnitudes of 8–9 MMS, cannot be predicted and could occur along any of tens of thousands of kilometres of plate margin, instantaneously.

The Kashmir earthquake

The 2005 Kashmir earthquake ranks as one of the most destructive in recent decades.

- At magnitude 7.6 and with a ground-shaking intensity of VII (severe), this was a large event.
- As with all earthquakes, speed of onset was very rapid so there was no chance of evacuating to a safe area.
- Damage was centred on Muzaffarabad but spread over an areal extent of more than 1000 km^2.
- Ground shaking lasted 30–45 seconds, but landslides triggered by the earthquake continued for some time, as did aftershocks up to magnitude 6.4.

Destruction in Kashmir was severe: 87,000 deaths, 2.8 million people displaced or made homeless, 17,000 schools destroyed or damaged and nearly 800 health centres destroyed. Numerous factors help explain the impacts, including poverty, poor building construction, time of day (many children were in school), geology and terrain (contributing to landslides) and isolation (making the rescue and relief effort difficult), but two hazard profile characteristics are also relevant:

1 Frequency: the previous major earthquake in Kashmir was in 1905, so there was no 'collective memory' of the risks and impacts of earthquakes in the region.

2 Spatial predictability: Kashmir is in a 'seismic gap', i.e. an area of known risk that had not experienced an earthquake for some time. There should have been education and risk awareness, which could have reduced the impacts.

Development and governance

The impacts of tectonic hazards are not only determined by level of development, but it is an important factor. Table 8 shows five recent earthquakes which all had the same 7.7 magnitude. There is some relationship between death toll and HDI (Human Development Index), such that lower HDI appears to suggest higher death tolls.

Table 8 Five magnitude 7.0 earthquakes compared

Date and location	Deaths	HDI
2013 Balochistan, Pakistan	825	0.54
2015 Nepal	9018	0.55
2010 Sumatra, Indonesia	711	0.68
2013 Khash, Iran	35	0.77
2007 Chile	2	0.83

Knowledge check 8

Which plate tectonic setting generates the most destructive, high-risk earthquakes?

Aftershocks occur in the hours, days and months after a primary earthquake and can be of high magnitude; they often number in the hundreds or thousands.

Exam tip

Physical process exam questions require you to use accurate and precise terminology.

However, other factors such as population density, duration of ground shaking, secondary hazards and response are also important. Generally, low level of development increases risk by increasing vulnerability, as shown in Table 9.

Table 9 Factors increasing and mitigating risk

Increasing risk	Mitigating risk
■ Population growth ■ Urbanisation and urban sprawl ■ Environmental degradation ■ Loss of community memory about hazards ■ Very young, or very old, population ■ Ageing, inadequate infrastructure ■ Greater reliance on power, water, communication systems	■ Warning and emergency-response systems ■ Economic wealth ■ Government disaster-assistance programmes ■ Insurance ■ Community initiatives ■ Scientific understanding ■ Hazard engineering

In some locations with very low levels of human development (HDI below 0.55), vulnerability is usually high because:

■ many people lack basic needs of sufficient water and food even in 'normal' times
■ much housing is informally constructed with no regard for hazard resilience
■ access to healthcare is poor, and disease and illness are common
■ education levels are lower, so hazard perception and risk awareness are low.

Many low-income groups lack a 'safety net' — either a personal one (savings, food stores) or a government one (social security, aid, free healthcare) — so have few resources after a disaster.

In rural Nepal, the area hit by the 2015 earthquake, 40% of families lived below the poverty line and more than 90% of people depended on subsistence farming. Of the rural population, 40% exhibited stunting as a result of malnutrition and only 20–40% of rural adults were literate.

Governance

Governance refers to the processes by which a country or region is run. Sometimes this is called 'public administration' and relates to how 'well run' a place is. 'Good governance' implies that national and local government are effective in keeping people safe, healthy and educated.

The effectiveness of governance varies enormously and has a significant impact on coping capacity and resilience in the event of a natural disaster. Table 10 explains the link between governance and vulnerability.

Knowledge check 9

Name a secondary hazard often associated with earthquakes in high terrain areas.

Table 10 Aspects of governance and disaster vulnerability

Meeting basic needs	Planning	Environmental management
When food supply, water supply and health needs are met, the population is physically more able to cope with disaster	Land-use planning can reduce risk by preventing housing on high-risk slopes, areas prone to liquefaction or areas within a volcanic hazard risk zone	Secondary hazards, such as landslides, can be made worse by deforestation. The right monitoring equipment can warn of some hazards, such as lahars
Preparedness	**Corruption**	**Openness**
Education and community preparation programmes raise awareness and teach people how to prepare, evacuate and act	Siphoning off money earmarked for hazard management, or 'kick-backs' and bribes to allow illegal or unsafe buildings increase vulnerability	Governments that are open, with a free press and media, can be held to account, increasing the likelihood that preparation and planning take place

> **Synoptic link**
>
> (P): Players in governance include local and national government and their agencies, which can be highly effective in terms of disaster response or ineffective and corrupt.

Most countries have national disaster management agencies, such as FEMA in the USA, which increase resilience to hazards and reduce the impacts of disasters. In the developing world these can be effective, such as PHIVOLCS in the Philippines, but they are often under-funded and under-resourced. Low-level corruption of local government officials is common in many developing countries, meaning that building codes are often ignored and construction is allowed in inappropriate places. This was widely blamed for the high death toll of 17,000 in the Izmit earthquake in Turkey in 1999.

Corruption refers to illegal practices, such as accepting bribes designed to influence decision making or paying people to stay silent about known problems.

Geographical factors

The nature of tectonic hazard impacts is influenced by a number of geographical factors. These include:

- population density: highly populated areas may be hard to evacuate, such as the area around Mt Vesuvius in Italy, and are likely to be hit harder by an earthquake
- degree of urbanisation: when cities are struck by major earthquakes, such as the 1995 Kobe earthquake in Japan or Haiti in 2010, death tolls can be high because of the concentration of at-risk people
- isolation and accessibility: often rural areas are hit less hard than urban areas by the initial impact of a tectonic disaster, but isolation and limited access can slow the rescue relief effort. The 2005 Kashmir earthquake is a good example.

Urban areas usually have more assets than rural areas. These include hospitals, emergency services, food stores and transport connections, which increase resilience and coping capacity compared with isolated rural places. However, high population density may mean more people are affected.

Knowledge check 10

Why are cities more vulnerable to earthquakes than rural areas?

Disaster context

No tectonic disaster can be separated from the wider local and national context within which it occurs. Table 11 illustrates this (see Table 6 on p. 16 for additional examples).

Table 11 Disaster context in Haiti and China compared

Developing country Haiti HDI = 0.48	Emerging country China HDI = 0.73
Port-au-Prince earthquake, 2010 160,000 deaths, 1.5 million homeless, 250,000 homes destroyed	Sichuan earthquake, 2008 69,000 deaths, 375,000 injured and economic costs of US$140 billion
Decades of corrupt, ineffective and brutal government left Haitian people hugely vulnerable because of slum housing, ineffectual water supply and endemic poverty. A post-earthquake cholera epidemic had infected 700,000 and killed 9000 by 2015	Economic losses in China were high, reflecting its development progress since 1990 (destroyed formal homes, businesses and infrastructure). The immediate response was rapid because the 2008 Beijing Olympic Games were only months away, so the Communist government mobilised the army and other responders rapidly

In developed countries major death tolls from tectonic hazards are rare. The 2011 Tohoku earthquake and tsunami in Japan was very much exceptional in terms of impacts. Countries such as Japan, the USA and Chile have:

- advanced and widespread insurance, allowing people to recover from disasters — at least in the long term
- government-run preparations such as Japan's Disaster Prevention Day on 1 September each year, as well as public education about risk, coping, response and evacuation
- sophisticated monitoring of volcanoes and, where possible, defences such as tsunami walls
- regulated local planning systems, which use land-use zoning and building codes to ensure buildings can withstand hazards and are not located in areas of unacceptable risk.

How successful is the management of tectonic hazards and disasters?

- There are important trends in tectonic disasters which need to be understood.
- Some locations are vulnerable to multiple hazards.
- Prediction and forecasting are important in managing disasters, but are not always possible.
- There are different models of disaster management but they are not universally applied.
- Different strategies can modify hazard events, vulnerability and loss, and reduce the impacts of events.

Tectonic disaster trends

The number of disasters and the impacts of disasters are not static. There is, however, a difference between two broad categories of natural hazard.

1 Hydrometeorological hazards, such as floods, storms, cyclones and drought, appear to have become more common over time, perhaps because of global warming and human environmental management issues such as deforestation.

2 Tectonic hazards, i.e. the events, have not increased or decreased over time. The number of events is broadly the same decade on decade.

Tectonic hazards and disasters are not the same, so even though the number of hazard events remains stable, the number of disasters has risen. Trends for all disaster types are shown in Figure 8 on p. 22.

Land-use zoning is a planning tool used to decide what type of buildings (residential, commercial, industrial or none) are allowed in particular locations.

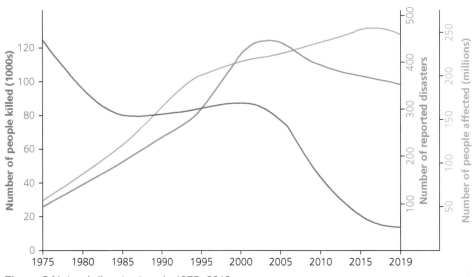

Figure 8 Natural disaster trends 1975–2019

Figure 8 shows three trends for all disasters.

1 Deaths have fallen over time because of better response management, preparation and, in some cases, prediction. Numbers of deaths have fallen especially since 2000, which may be due to vastly improved mobile communications to warn people of disasters.

2 The number of reported disasters increased then stabilised as improvements in data coverage and the accuracy of databases increased. Decades ago many disasters in isolated areas simply went unreported. Most recently, numbers of reported disasters have fallen, suggesting fewer hazard events are becoming disasters.

3 The number of people affected by disasters has recently stabilised but populations continue to grow and more people live in risky locations.

Trends in tectonic hazards can be summarised as follows.

■ There has been no change in the number of earthquake disasters since 1980, which varies between about 15 and 40 each year.

■ Earthquake deaths are very variable: there were fewer than 10,000 deaths worldwide annually from 2012 to 2018, but more than 200,000 in 2004 and 2010. Overall, there are fewer earthquake deaths than there were 30–40 years ago but the impact of single, **mega-disasters** skews the data.

■ The trend for earthquake economic losses is upwards, averaging about US$20–40 billion per year, but once again this is affected by a few very large events.

Economic losses from tectonic disasters continue to rise. More people, who are more affluent, have more property to lose. This is increasingly true in emerging countries as well as developed ones.

Volcanic disasters are much less frequent than earthquake ones and deaths from eruptions are now rare. The last time an eruption killed more than 1000 people was in Cameroon in 1986 (Lake Nyos) and only eight eruptions since 1980 have killed more than 100 people. However, numbers affected can still be very large because of

Exam tip

You can sketch graphs, such as Figure 8, as part of your exam answer.

Mega-disasters are high-magnitude, high-impact, infrequent disasters that affect multiple countries directly or indirectly so their impacts are regional or even global.

the mass evacuation of people around an erupting volcano, e.g. 130,000 affected by the eruption of Mt Merapi in Indonesia in 2010 but only 300 deaths.

Mega-disasters and multiple hazard zones

Very large tectonic disasters account for most deaths. There were about 300 deadly earthquakes between 2005 and 2019. Of the 440,000 people killed, 418,000 were killed by just five disasters. Three of these — Kashmir 2005, Sichuan 2008 and Nepal 2015 — are in the same tectonic location, i.e. the Himalaya collision zone. These three disasters accounted for 38% of all earthquake deaths between 2005 and 2019, and the 2010 Haiti earthquake accounted for about another 50%.

In recent years three examples of what might be called 'mega-disasters' have occurred (Table 12). Although rare, these are characterised by impacts extending beyond the country immediately affected. The 2011 tsunami in Japan showed how the globalised, inter-dependent world economy could be affected by the economic and human impacts of disasters. In addition, the accompanying nuclear meltdown disaster at Fukushima was a catalyst in Germany abandoning its nuclear electricity programme.

Table 12 Mega-disaster impacts

	Number of countries affected	Impacts
2004 Asian tsunami	14 countries surrounding the Indian Ocean	Economic losses and deaths in Indonesia, Thailand, Sri Lanka and Somalia among others made this disaster one of the largest ever in terms of areal extent
2011 Japanese tsunami	Only Japan directly, but the economic impacts had global consequences	Disruption to ports, factories and power supplies had impacts for the global car-production supply chain and those of Boeing jets and semiconductors used in modern electronics
2010 Eyjafjallajökull eruption	Over 20 European countries were affected by total or partial closure of their airspace	The ash cloud from the Eyjafjallajökull eruption had a disruptive effect on air travel because of the dangers of jet engines ingesting ash: over 100,000 cancelled flights costing over £1 billion in losses

A number of locations are **multiple hazard zones** (Figure 9). These include California, the Philippines, Indonesia and Japan. These locations:

- are tectonically active and so earthquakes (and often eruptions) are common
- are geologically young with unstable mountain zones prone to landslides
- are often on major storm tracks either in the mid-latitudes or on tropical cyclone tracks
- may suffer from global climate perturbations such as El Niño/La Niña.

Famously, during the 1991 eruption of Mount Pinatubo in the Philippines the area was struck by Typhoon Yunga. Heavy rainfall from the typhoon mobilised volcanic ash into destructive lahars. This shows how linked hydrometeorological hazards can contribute to tectonic disasters. This eruption could have been significantly worse in terms of impact but it was successfully predicted and evacuation limited the death toll

Knowledge check 11

Worldwide, which disaster impact has a rising trend: deaths, number of reported disasters or numbers affected?

Exam tip

Learn some key facts and figures about named disaster events — they will make your answers more convincing.

Knowledge check 12

State two characteristics of a mega-disaster.

Multiple hazard zones are places where two or more natural hazards occur, and in some cases can interact to produce complex disasters.

to about 850. In many earthquake-prone areas, landslides can be triggered by heavy rain on slopes previously weakened by earthquake tremors.

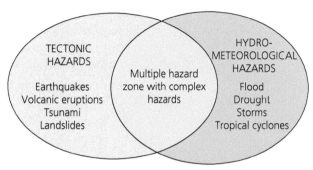

Figure 9 Multiple hazard zones

Prediction and forecasting

Prediction is the holy grail of hazard management but it is not always possible, as outlined in Table 13. Despite decades of scientific research earthquakes cannot be predicted. However, the minimal death toll from volcanic eruptions (despite 60–80 eruptions worldwide per year) can be mainly attributed to vastly improved prediction of these events.

Prediction means knowing when, and where, a natural hazard will strike on a spatial and temporal scale that can be acted on meaningfully in terms of evacuation.

Forecasting is much less precise than prediction. It provides a 'percentage chance' of a hazard occurring, e.g. a 25% chance of a magnitude 7.0 earthquake in the next 20 years.

Table 13 Predicting tectonic hazards

Hazard type	Prediction?	Further details
Earthquakes	No	■ Only areas at high risk can be identified (risk **forecasting**), plus areas that are likely to suffer severe ground shaking and liquefaction; this can be used for land-use zoning purposes ■ 'Seismic gaps', i.e. areas that have not experienced an earthquake for some time and are 'overdue', can point to areas of especially high risk
Volcanic eruptions	Yes	■ Sophisticated monitoring equipment on volcanoes can measure changes as magma chambers fill and eruption nears ■ Tiltmeters and strain meters record volcanoes 'bulging' as magma rises and seismometers record minor earthquakes indicating magma movement ■ Gas spectrometers analyse gas emissions which can point to increased eruption likelihood
Tsunami	Partly	■ An earthquake-induced tsunami cannot be predicted ■ However, seismometers can tell an earthquake has occurred and locate it, then ocean monitoring equipment can detect tsunami in the open sea ■ This information can be relayed to coastal areas, which can be evacuated

Synoptic link

(P): Scientists are key players in prediction and forecasting, investigating and understanding physical processes, modelling forecasts using supercomputers and developing monitoring technology for volcanic eruptions and tsunami.

Exam tip

Make sure you use the words 'prediction' and 'forecasting' carefully as they have very different meanings.

Prediction of tsunami and eruptions depends on technology which has to be in place, operational and linked to warning, dissemination and evacuation systems. Tsunami monitoring equipment was not present in the Indian Ocean in 2004 so there was no way of warning people on distant coasts — despite there being many hours in which to do so.

In many developing countries, volcano monitoring and tsunami warning may not be as good as they could be because of the cost of the technology. Also, it may be more difficult to reach isolated, rural locations with effective warnings.

Knowledge check 13

Can earthquakes be predicted?

Hazard management

Prediction, when possible, is a vital part of attempts to manage the impacts of natural disasters. However, it is not the only approach. The hazard management cycle shown in Figure 10 (sometimes called the disaster management cycle) illustrates the different stages of managing hazards in an attempt to reduce the scale of a disaster. It is important to see this as a cycle, with one disaster event informing preparation for the next.

PREPAREDNESS Community education and resilience building including how to act before, during and after a disaster, prediction, warning, and evacuation technology and systems		RESPONSE Immediate help in the form of rescue to save lives and aid to keep people alive, emergency shelter, food and water
MITIGATION Acting to reduce the scale of the next disaster: land-use zoning, hazard-resistant buildings and infrastructure		RECOVERY Rebuilding infrastructure and services, rehabilitating injured (physically and mentally) peoples and their lives

Figure 10 The hazard management cycle

The 'recovery' stage of the hazard management cycle might be thought of as the 'returning to normal' stage. This can happen after a few months but in some cases it takes years. The recovery stage depends on:

- the magnitude of the disaster — bigger means longer
- development level — lower means longer, as poorer people are more severely affected
- governance, because well-governed places will divert resources more effectively to recovery efforts
- external help, i.e. aid and financing to help the recovery effort.

The importance of recovery can be seen on Park's model: the disaster response curve. This well-known model provides a simple visual illustration of the impact of a disaster. It can be used to compare different hazard events as shown by curves A–C on Figure 11.

Synoptic link

(P): Emergency planners are key players in disaster management. The quality of these planners and the funding they receive from government are critical in terms of minimising impacts.

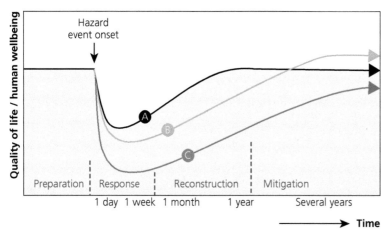

Figure 11 Park's model
Source: after Park (1991)

- Curve A shows a disaster with a relatively small impact on quality of life and a short response phase. Quality of life begins to improve quickly and returns to normal within a few months.
- Curve B shows quality of life is impacted more than in A and reconstruction takes longer, but mitigation improves quality of life, meaning the community is better prepared for the next hazard.
- Curve C shows a disaster with a major impact on quality of life and a slow reconstruction phase; even years later quality of life has not returned to levels before the disaster.

It could be argued that the 2010 earthquake in Haiti fits profile C because the devastation was still not 'fixed' by 2015 and incomes, health, housing and food supply remained worse than pre-disaster. Developed countries are more likely to correspond to profiles A or B, developing ones to C.

Disaster modification

Disasters can be managed by modifying impacts. This can be done in three ways (Figure 12): by modifying the event, modifying vulnerability or modifying loss.

Exam tip

Be prepared to compare and contrast examples in the exam, using concepts such as Park's model as a framework for your answer.

Knowledge check 14

Which stage of the hazard management cycle would involve rebuilding schools, hospitals and businesses?

Event modification is the most desirable type of management but it is not always possible. Loss modification implies that a disaster has occurred and caused damage to people and property: this is the least desirable form of management.

Figure 12 Disaster modification

Event modification (Table 14) relies on technology and planning systems and can be high cost. It is less likely to be used in developing and emerging countries, although low-cost examples of **hazard resistant design** are possible to build.

Table 14 Modify the event

Type of modification and example	Advantages	Disadvantages
Land-use zoning: preventing building on low-lying coasts (tsunami), close to volcanoes and areas of high ground-shaking and liquefaction risk	■ Low cost ■ Removes people from high-risk areas	■ Prevents economic development on some high-value land, e.g. coastal tourism ■ Requires strict, and enforced, planning rules
Aseismic buildings: cross-bracing, counter-weights and deep foundations prevent earthquake damage	■ Widely used technology can prevent collapse ■ Protects both people and property	■ High costs for tall/large structures ■ Older buildings and low-income homes are rarely protected (Figure 13)
Tsunami defences: tsunami sea walls and breakwaters prevent waves travelling inland	■ Dramatically reduce damage ■ Provide a sense of security	■ Can be overtopped ■ Very high cost ■ Ugly and restrict use/development at the coast
Lava diversion: channels, barriers and water cooling used to divert and/or slow lava	■ Diverts the lava out of harm's way ■ Relatively low cost	■ Only works for low VEI basaltic lava ■ The majority of 'killer' volcanoes are not of this type

Hazard resistant design involves constructing buildings and infrastructure that are strong enough to resist tectonic hazards. In the case of earthquakes these are called aseismic buildings.

'Quake-proof' house

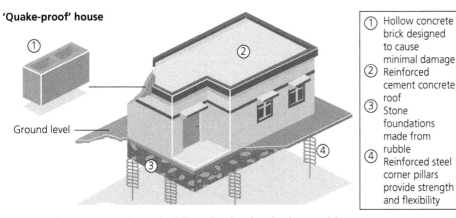

① Hollow concrete brick designed to cause minimal damage
② Reinforced cement concrete roof
③ Stone foundations made from rubble
④ Reinforced steel corner pillars provide strength and flexibility

Ground level

Figure 13 Low-cost aseismic buildings for the developing world

Modifying vulnerability (Table 15) means increasing the resilience of a community to increase their capacity to cope. In many cases, prediction, warning and evacuation are used to move people out of harm's way. Predictions need to be accurate or there is a risk that 'cry wolf syndrome' will reduce the effectiveness of warnings.

Cry wolf syndrome occurs when predictions (and evacuation) prove to be wrong, so that people are less likely to believe the next prediction and warning and therefore fail to evacuate.

Table 15 Modify the vulnerability

Type of modification and example	Advantages	Disadvantages
Hi-tech scientific monitoring: used to monitor volcano behaviour and predict eruptions	▪ In most cases, predicting an eruption is possible ▪ Warnings and evacuation save lives	▪ Costly, so not all developing world volcanoes are monitored ▪ May suffer from 'cry wolf syndrome' ▪ Does not prevent property damage
Community preparedness and education: earthquake kits and preparation days, education in schools	▪ Low cost; often implemented by NGOs ▪ Can save lives through small actions	▪ Does not prevent property damage ▪ Harder to implement in isolated rural areas
Adaptation: moving out of harm's way and relocating to a safe area	▪ Would save both lives and property	▪ High population densities prevent it ▪ Disrupts people's traditional homes and traditions

Earthquake kits are boxes of essential household supplies (water, food, battery-powered radio, blankets) kept in a safe place at home to be used in the days following an earthquake.

Synoptic link

(P): Civil engineers are the players responsible for building tsunami walls and aseismic buildings. Their expertise can directly save lives, although it is governments that generally fund such defences.

Exam tip

Remember that different types of modification are more applicable to some tectonic hazards than others, and cannot always be used in the developing world because of cost restrictions.

Loss modification could be described as 'picking up the pieces' after a disaster has occurred (Table 16). If event and vulnerability modification has also been used then the losses should be quite small. However, in the case of developing countries, loss modification is often the main management strategy. This was the case after the 2010 Haiti earthquake and the 2004 Indian Ocean tsunami. In these cases management should be considered as having failed to protect people.

Knowledge check 15

What type of disaster modification is least likely to save lives?

Table 16 Modify the loss

Type of modification and example	Advantages	Disadvantages
Short-term emergency aid: search and rescue followed by emergency food, water and shelter	▪ Reduces death toll by saving lives and keeping people alive until longer-term help arrives	▪ High costs and technical difficulties in isolated areas ▪ Emergency services are limited and poorly equipped in developing countries
Long-term aid: reconstruction plans to rebuild an area and possibly improve resilience	▪ Reconstruction can 'build in' resilience through land-use planning and better construction methods	▪ Very high costs ▪ Needs are quickly forgotten by the media after the initial disaster
Insurance: compensation given to people to replace their losses	▪ Allows people to recover economically, by paying for reconstruction	▪ Does not save lives ▪ Few people in the developing world have insurance

Synoptic link

(P): Major insurance companies are the players that help some people recover by covering their losses; however, in the developing world NGOs often step in to help the uninsured and poorest sections of society. Even in developed countries not everyone is insured against the impacts of disasters.

Summary

- The distribution of earthquakes, volcanoes and tsunami is related to plate margins (divergent, convergent and conservative) and explained by the theory of plate tectonics.
- Tectonic processes result in a number of hazard types, including crustal fracturing, ground shaking, liquefaction and landslides, as well as a variety of volcanic hazards and tsunami.
- Natural hazards and disasters are not the same: disasters happen when a vulnerable population with low resilience experiences the negative impacts of natural hazards.
- Social, economic and environmental impacts vary by level of development as well as type of hazard process.
- Magnitude, frequency and the hazard profile of events are all important in terms of understanding variation in impacts.
- The urban, economic, demographic, political governance and social context of an area are all important in understanding vulnerability and community resilience.
- Trends show that while deaths and numbers affected by disasters are falling, economic damage is rising.
- Mega-disasters and multiple hazard zones are particularly important in terms of large-scale, sometimes global, impacts.
- There is variation in the extent to which tectonic hazards can be predicted.
- Models such as Park's model and the hazard management cycle can be useful in terms of understanding short- and long-term impacts, response, recovery and mitigation.
- Tectonic hazards can be managed by modifying the event, vulnerability or loss — each approach has advantages and disadvantages.

■ Landscape systems, processes and change

Glaciated landscapes and change

How has climate change influenced the formation of glaciated landscapes over time?

- The Earth's climate has fluctuated between warm interglacial and cold glacial conditions many times in the last few millions of years, on different timescales and with different causes.
- Ice cover today is very different from ice cover that was present in the past.
- Periglacial areas, at the edges of areas of ice cover on land, have also seen their distribution change over time.

Causes of climate change

Earth's **climate** has changed significantly in the past. Throughout Earth's history there have been two dominant climates:

1 Icehouse conditions, when ice cover has been present across large areas.

2 Greenhouse conditions, when the climate has been much warmer and largely ice free.

For the last 30 million years, icehouse conditions have prevailed. The last 2.6 million years are referred to as the Quaternary geological period. This is subdivided into two epochs:

1 the Pleistocene, from 2.6 million to 12,000 years ago

2 the Holocene, from 12,000 years ago to the present day.

The Pleistocene saw more than 20 major climate fluctuations where Earth's climate flipped between:

- interglacial periods, with warm temperatures similar to Earth's climate today
- glacial periods, with much colder temperatures compared with today and extensive ice cover.

Each cold–warm cycle lasted around 100,000 years.

Natural causes

The accepted main cause of climate change in the Pleistocene is changes in the Earth's orbit around the sun (orbital changes or astronomical forcing). This is based on Milankovitch cycle theory developed by Milutin Milankovitch in 1924. Milankovitch argued that the surface temperature of the Earth changes over time because the Earth's orbit and axis alignment vary over time. These variations lead to changes in the amount and distribution of solar radiation received by Earth from the sun.

Climate refers to the 30-year average temperature and precipitation conditions for an area. It is distinct from weather, which refers to day-to-day changes in conditions.

Exam tip

Learn the approximate dates of geological time periods such as the Holocene and Pleistocene.

- On a timescale of 100,000 years, Earth's orbit changes from circular to **elliptical** and back again (called orbital eccentricity). This changes the amount of radiation received from the sun.
- On a timescale of 41,000 years Earth's axis tilts from 21.5° to 24.5° and back again (axial tilt). This changes the seasonality of Earth's climate. The smaller the tilt, the smaller the difference between summer and winter.
- Lastly, on a 22,000-year timescale, Earth's axis 'wobbles' and this changes the point in the year that the Earth is closest to the sun (axial precession).

Figure 14 shows the combined effects of these three cycles. Milankovitch's theory is supported by the fact that glacial periods have occurred at regular 100,000-year intervals.

An **elliptical** orbit is one shaped like a rugby ball rather than a football (a circular orbit).

Knowledge check 16

Which of the three orbital changes has the largest impact on climate change?

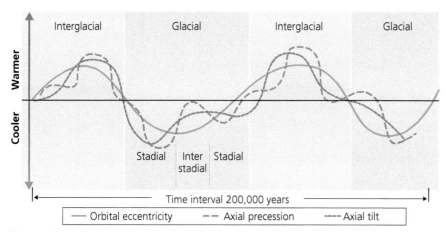

Figure 14 The combined effect of Milankovitch cycles

The impact of orbital changes on solar radiation amount and distribution is small. In total, it can probably change global temperature by ±0.5°C. The evidence of past climate change from ice cores and other evidence suggests that glacial periods were 5–7°C colder than interglacials. Scientists think that orbital changes may have been just enough to trigger a major global climate change, but climate **feedback mechanisms** are needed to sustain it.

- An example of positive feedback is snow and ice cover. Small increases in snow and ice dramatically raise surface **albedo** and reflect solar energy back into space. This contributes to further cooling, which might encourage further snowfall, and therefore a cycle of cooling. This may be how a 0.5°C cooling caused by orbital changes could be amplified into a 5°C global cooling.
- An example of negative feedback is cloud cover. As warming occurs, more evaporation will occur and this may increase global cloud cover. Increasingly cloudier skies could reflect more solar energy back into space and diminish the effect of the warming.

Solar output

The amount of energy emitted by the sun varies as a result of **sunspots**. The effect of sunspots is to blast more solar radiation towards the Earth. There is a well-known 11-year sunspot cycle, as well as longer cycles. The total variation in solar radiation

Feedback mechanisms can either amplify a small change and make it larger (positive feedback) or diminish the change and make it smaller (negative feedback).

Albedo means the reflectivity of a surface. Snow and ice have a high albedo, reflecting back heat energy, whereas oceans and forests have a low albedo and absorb heat energy.

Sunspots are dark spots that appear on the sun's surface, caused by intense magnetic storms.

caused by sunspots is about 0.1%. Sunspots have been recorded for around 2000 years, and there is a good record for around 400 years.

A long period with almost no sunspots, known as the Maunder Minimum, occurred between 1645 and 1715, and this is often linked to the Little Ice Age (Figure 15). Similarly, the Medieval Warm Period has been linked to more intense sunspot activity, although it is unclear whether the Medieval Warm Period was a global event. The Little Ice Age climate cooling event:

- lasted from 1450 to 1850
- experienced especially cold periods within that time period, at around 1660, 1770 and 1850
- had average temperatures that were 0.5–1°C colder than the 20th century
- caused rivers and lakes to freeze more regularly and sea ice in the Arctic to be more extensive
- made the growing season for farmers shorter and less reliable, leading to food shortages.

Volcanic eruptions

Major volcanic eruptions eject material into the stratosphere where high-level winds distribute it around the globe.

- Volcanoes eject huge volumes of ash, sulphur dioxide, water vapour and carbon dioxide.
- High in the atmosphere, sulphur dioxide forms a haze of sulphate aerosols, which reduces the amount of sunlight received at the surface.
- Tambora, Indonesia, ejected 200 million tonnes of sulphur dioxide in 1815 and in 1991 Mt Pinatubo ejected 17 million tonnes.

Tambora led to the 'year without a summer' in 1816 as global temperatures dipped by 0.4–0.7°C. Temperature dropped by 0.6°C following the Pinatubo eruption. These are short-lived effects, as the sulphate aerosols persist for only 2–3 years.

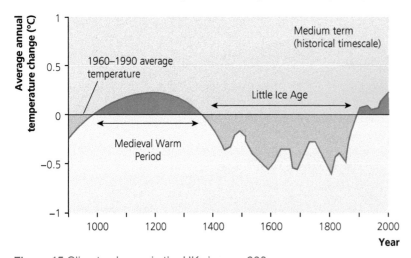

Figure 15 Climate change in the UK since AD 900

Exam tip

Take care not to overestimate the initial impact on the climate of sunspots and orbital changes, which is small (but may be amplified by feedback mechanisms).

Knowledge check 17

Which climate change episode is linked to the Maunder Minimum?

Loch Lomond stadial

During the Pleistocene, at the end of the last glacial period (called the Devensian) when ice sheets were melting, there was a short but severe return to very cold conditions in the North Atlantic:

■ This Loch Lomond **stadial** event began 12,700 years ago and 'ice age' conditions returned. Glaciers began to grow in the Scottish Highlands. Temperatures in the British Isles ranged from 10°C in summer to −20°C in winter. After about 1300 years, temperatures suddenly rose and have been warm ever since (see Figure 16).

■ The most probable explanation is a sudden influx of cold freshwater to the North Atlantic caused by the melting ice sheets. This disrupted the **Thermohaline Circulation** in the North Atlantic and this sudden influx ended only when the supply of glacial meltwater ran out. This event tells us that natural climate changes can occur very rapidly.

A **stadial** is a short cold period within a warmer interglacial. The term interstadial refers to a brief warming during a cold glacial period.

The **Thermohaline Circulation** is a system of interconnected ocean currents that helps redistribute heat around the Earth, which helps even out temperature differences between the equator and the poles.

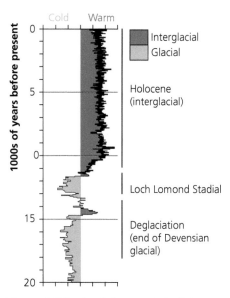

Figure 16 The Loch Lomond stadial

Knowledge check 18

What was the last glacial period called?

Exam tip

It can be useful to sketch graphs, such as Figures 15 and 16, in the exam as it may save time.

Past and present ice cover

Ice cover found on planet Earth is referred to as the **cryosphere**. Ice is found in a wide range of different environments that broadly divide into:

■ high latitude: ice found within the Arctic and Antarctic Circles, more than 65 degrees of latitude north and south

■ high altitude: ice found in mountain ranges, which can occur at any latitude.

The **cryosphere** is the discontinuous layer of ice covering parts of the Earth's land and ocean surface, as well as ice found underground.

There are a number of different types of ice mass, shown in Table 17.

Table 17 Types of ice mass

Ice mass type	Description	Area (km²)	Examples
Ice sheet	Large, dome-shaped ice masses that completely submerge underlying topography and are many kilometres thick	More than 50,000	Antarctica and Greenland
Ice cap	A smaller version of an ice sheet, with its high point corresponding to the high point of submerged mountains beneath the ice	Less than 50,000	Austfonna on Svalbard
Ice field	Ice covering a mountain plateau, but not thick enough to bury all of the topography or extend beyond the high-altitude area	Less than 50,000	Garden of Eden ice field, New Zealand
Valley glacier	A long tongue of ice confined between valley sides and terminating in the valley or at the sea; outlet glaciers drain ice caps and ice sheets	5–1500	Arolla Glacier (Switzerland/Italy)
Cirque glacier	Smaller glacier occupying a hollow in a mountainside; circular in shape (also called a corrie glacier)	0.5–10	Hodges Glacier, Grytviken, South Georgia

> **Exam tip**
> Learn definitions of each of the types of ice mass, as well as named examples of them.

Ice masses can be constrained or unconstrained. Ice sheets and ice caps blanket the topography beneath them (unconstrained) whereas ice fields and glacier edges butt up against valley walls and mountain sides (constrained). Ice is also found in the sea, as ice shelves. This floating sea ice generates icebergs when wave action carves off chunks of the ice.

Ice masses can also be classified by **thermal regime**.

- Warm-based (found in temperate locations) glaciers are not frozen to the surface they sit on, i.e. the bedrock. At their base there is a layer of water, allowing the base of the ice to slide.
- Cold-based glaciers (found in polar high latitudes) have very cold surface temperatures of −20°C to −30°C and their base is frozen to the bedrock beneath. Neither the pressure of the ice above nor geothermal energy from below is enough to melt the ice base.
- Polythermal glaciers have a cold-based upper portion at high altitude but a warm-based lower portion at lower altitude.

> **Thermal regime** refers to the temperature at the base of an ice mass. It determines whether there is ice or water present.

> **Knowledge check 19**
> What is the thermal regime of glaciers that slide on a layer of water at their base?

Distribution of ice cover

The cryosphere plays a key role in a number of global systems.

- It is a key component of the hydrological cycle, with about 69% of the world's fresh water locked up in ice. Ice captures water from the atmosphere and glacier melt releases this freshwater into river systems.
- Surface ice has a significant cooling effect on the planet because of ice's high albedo. This reflects solar radiation back into space, cooling polar regions and regulating Earth's climate.
- Polar regions, especially **permafrost** areas, are significant global carbon stores, as biological carbon is locked away in frozen ground, preventing bacterial decay. This helps regulate the level of carbon dioxide in the atmosphere.

> **Permafrost** is soil, sediment or rock that has been frozen for two or more years.

Today ice cover can be found on land and in the ocean, as shown in Table 18.

Table 18 The extent of ice on land and in the oceans

Ice on land	Percentage of global land surface
Antarctic ice sheet	8.3
Greenland ice sheet	1.2
Glaciers, ice caps and ice fields	0.5
Permafrost	9–12
Seasonal permafrost	33
Ice in the ocean	Percentage of global sea surface
Antarctic ice shelves	0.5
Sea ice (winter maximum)	9.1
Sub-sea permafrost	0.8

The vast majority of ice is in either permafrost or found in Antarctica. Glaciers, ice caps and ice fields are a very small proportion of the total ice cover. Today there are around 170,000 valley glaciers and numerous ice caps in the world but most are found in a small number of locations.

At the height of the last glacial period in the Pleistocene, ice covered about 32% of the total land area, compared with about 10% today (Figure 17). Ice sheets covered a huge area of North America, northern Europe and north and central Asia as well as the Andes. In addition:

- sea ice extent was much larger, extending as far south as Japan and as far north as Argentina
- sea level was much lower because so much water was locked up in land-based ice.

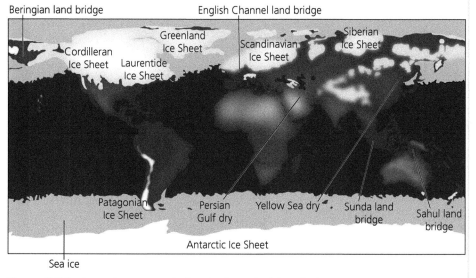

Figure 17 Maximum ice extent during the last glacial period, 30,000 years ago

There is extensive evidence of past high-altitude ice cover in the form of **relict landscapes** from the Pleistocene. The UK has many upland areas that were formerly glaciated. These include the Highlands and Grampians, the Lake District and Snowdonia.

Knowledge check 20

Which is the largest single ice mass on Earth's surface?

Relict landscapes are wholly or partly preserved landscapes, which show physical landform evidence for very different climate, environmental and physical process conditions in the past.

Periglacial environments

Locations at the edge of ice-covered areas are called periglacial environments. In these places the defining feature of the landscape is that the ground is frozen — called permafrost. Periglacial environments are found:

- at high latitudes, especially in the Arctic. They are less common in the Southern Hemisphere because of the absence of landmasses between 60°S and 70°S
- at high altitude because temperatures fall by 1°C for every 100 m of height, e.g. the Himalayan plateau
- in continental interiors, away from the moderating influence of the sea, e.g. Siberia.

Figure 18 shows ice cover and permafrost extent in the Northern Hemisphere. It should be noted that permafrost extends out under the sea bed in the Arctic Ocean.

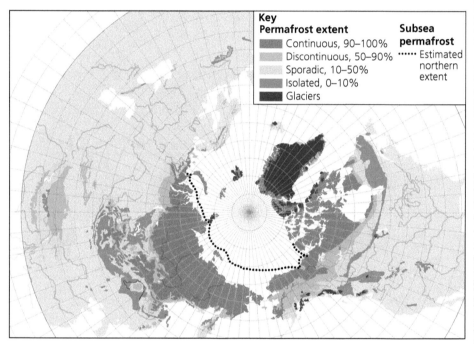

Figure 18 The extent of permafrost in the Northern Hemisphere

Source: Permafrost Laboratory, Geophysical Institute, University of Alaska, 2007

There are different types of permafrost:

- continuous permafrost (90%+ of the ground surface is underlain by permafrost) in areas where mean annual soil temperature is below −5°C
- discontinuous permafrost (50–90% of the ground surface) found in areas where the mean annual temperature is below −1°C
- sporadic permafrost (less than 10–50% of the ground surface) exists as patches, often in areas shaded from direct sunlight or where soils are more easily frozen, and is termed 'isolated' when the percentage cover falls below 10%.

Freezing extends downwards to the point where **geothermal heat** from below prevents freezing. This can be deeper than 1000 m. Permafrost has a summer seasonally active layer. This is the upper layer which seasonally melts. The depth

Geothermal heat is heat conducted from within the Earth towards the surface.

varies from a few centimetres to several metres, depending on the average summer temperature.

In the past, the area of permafrost was much more extensive. At the height of the last glacial period, permafrost covered most of southern England and the landscape was one of **tundra**.

Periglacial processes

Periglacial areas are affected by a number of physical processes, many of which are highly seasonal and dependent on annual cycles of freezing and thawing. These processes contribute to a distinctive periglacial landscape and landforms in places such as northern Russia and northern Canada (Table 19).

Table 19 Periglacial processes and landforms

Process	Explanation	Related landforms
Nivation	Seasonal snow collects in shallow depressions and moves slowly downslope	Nivation hollows: weathering, erosion and meltwater flow caused by the snow's movement deepen the depression into a bowl-shaped hollow
Frost heave	The upward swelling of soil due to the growth of ice lenses within the soil. The ice lenses are fed by capillary action or groundwater flow from below. Frost heave also forms **patterned ground**	Pingos: these are circular, soil-covered mounds up to 70 m high which have an ice core, usually fed by a groundwater source from below allowing the ice core to swell upward, forcing the soil mound upward
Thermal contraction	The shrinkage of the ground surface due to extreme low temperatures, creating cracks in the surface with a regular, interlocking pattern of polygons (patterned ground)	Ice wedge polygons: ice wedges form when spring meltwater pools in thermal contraction cracks, only to refreeze when it hits permafrost below. Over many seasons the ice wedge grows in thickness
Freeze–thaw weathering	On freezing, water expands by 9% in volume. Repeated cycles of freezing and thawing of water in cracks in rock splits rocks into progressively smaller blocks	Blockfields and scree: frost-shattering, angular rock fragments can form on a flat surface as a blockfield, or at the base of a slope, forming scree
Solifluction	A type of mass movement, or soil creep, found on low-angle slopes. Saturated soil of the active layer flows very slowly downslope in summer but freezes in the winter	Solifluction terraces and lobes: solifluction produces vegetated lobes or terraces giving a slope a 'stepped' profile
Wind erosion and deposition	Periglacial areas often have strong winds and these can erode the finely ground rock debris produced by glaciers and ice caps	Loess fields: extensive areas of wind-blown glacial sediment found at the margins of ice

A final feature of periglacial areas is the extreme seasonality of their **river regimes**. Rivers have no discharge in winter because of freezing conditions, but this can change rapidly in summer as seasonal melting begins, leading to extensive (but short-lived) meltwater erosion on river banks.

The **tundra** is the ecosystem or biome found in most periglacial areas (alpine tundra if at high altitude). It has no trees, low rainfall and consists mostly of low-growing grassland plants adapted to very cold and dry conditions.

Exam tip

You need to be able to recognise the landforms in Table 19 in photographs and diagrams, as well as explain their formation.

Knowledge check 21

Which process, in periglacial areas, breaks rocks down into smaller angular fragments?

Patterned ground is a unique feature of periglacial areas consisting of polygon, net and stripe patterns on the ground surface, made of either stones or ice wedges.

A **river regime** is the change in discharge experienced by a river over the course of a year, its seasonal flow variation.

What processes operate within glacier systems?

- Glaciers operate as systems, and the concept of mass balance is important in understanding the system.
- Ice masses move in different ways and at different rates.
- The action of ice creates unique landscapes as a result of erosion, entrainment, transport and deposition processes.

Glaciers as systems

Glacial ice begins to form when snowfall from one winter lasts until the next winter. Further snowfall begins the process of building multiple layers of ice. The basic process is one of compression as the weight of new snow above compresses old snow below.

- **Firn** forms after about 2 years, as snow recrystallises into sugar grain-sized crystals with small pore spaces between them.
- Ice forms slowly, over 20–30 years in some cases, as further compression and recrystallisation take place.

In very cold places such as Antarctica, the transformation of snow to ice may take hundreds of years.

Ice masses are a system of inputs (snow), transfers (ice movement) and outputs (**meltwater**). The snow and meltwater can be gained (accumulation) and lost (ablation) in a number of ways (Table 20).

Table 20 Ice masses as systems

Accumulation	Ice movement	Ablation
• Direct snowfall onto the ice mass • Avalanches carrying snow onto an ice mass from upslope • Windblown snow being moved from another area and deposited onto an ice mass	→	• Melting of ice at the margins of the ice mass • Calving of ice from the margin on an ice mass where it meets the sea • Evaporation and sublimation from the ice mass surface • Avalanches from the ice mass itself

For an ice mass to form, accumulation has to be greater than ablation. This occurs:
- in a high-altitude environment, such as a mountain valley or depression in the landscape
- in a high-latitude location, where average annual temperatures are very low.

As a mass of ice accumulates it will begin to flow away from the zone of accumulation. Eventually the ice mass will reach an area where temperatures are higher (a lower altitude and/or latitude) and ablation will exceed accumulation.

Glacier mass balance

The balance between ice mass inputs and outputs is called mass balance. This is shown diagrammatically in Figure 19. In a valley glacier such as the one in Figure 19,

Firn is the initial stage of ice formation, with a density of 550–830 kg/m³ (compared with 500 kg/m³ for snow or névé; ice has a density of about 900 kg/m³).

Exam tip

Many glacial processes need to be explained as a step-by-step, logical sequence, using precise terminology.

Knowledge check 22

What is 'firn'?

Meltwater is the flow of water from streams and rivers at the snout of a glacier or edge of a larger ice mass.

Knowledge check 23

What term is used to describe the build-up of snow on an ice mass, usually in the winter?

the mass balance is in equilibrium if accumulation and ablation cancel each other out over the course of a year. Figure 20 shows that over the course of a year there is a period of net accumulation (winter) and net ablation (summer), but over the whole year the system is in balance.

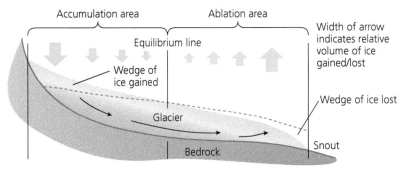

Figure 19 Glacier mass balance

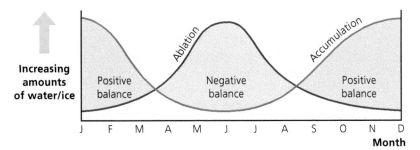

Figure 20 The annual ice budget of a glacier in the Northern Hemisphere

Ice masses with net ablation will shrink over time, whereas ones with net accumulation will grow. Feedback mechanisms can have an impact on mass balance.

- A small increase in snowfall caused by a lowering of temperature can increase the albedo of the ice surface, contributing to further cooling and more accumulation (positive feedback).
- It is possible that higher temperatures in some areas can increase evaporation and therefore precipitation (snowfall) so more snow accumulation balances the losses from ablation in higher temperatures (negative feedback).

Figure 20 shows how mass balance changes on the timescale of a year. Mass balance also changes on longer timescales as ice masses respond to climate trends. Many Alpine glaciers grew during the Little Ice Age and their snouts extended lower down alpine valleys. Mass balance was often close to **equilibrium** (= 0) in the 1980s, but there has been an accelerating loss of ice since. This change is usually ascribed to the impact of global warming.

Glacier movement

Ice moves because of gravity. Although solid, ice can deform before it fractures so is often described as 'flowing', rather like a fluid. The velocity of moving ice varies from about 0.5 m per year at the centre of cold-based ice sheets to up to 10 m per *day* at the snouts of some glaciers.

Exam tip

It is useful to be able to draw a quick sketch of a mass balance diagram, such as Figure 19, for exam questions on this topic.

Equilibrium means being in balance, i.e. equal amounts of accumulation and ablation over a year or other timescale.

Knowledge check 24

What is thought to be the main cause of the increased negative mass balance of glaciers seen worldwide in the last few decades?

In general, cold-based polar ice masses move very slowly and warm-based temperate high-altitude glaciers flow faster. However, some ice streams that drain polar ice sheets and caps move much faster than their parent ice mass.

There are several different mechanisms of ice movement, outlined in Table 21.

Table 21 How glacial ice moves

Basal slip	Ice slides across bedrock/sediment substrate because of a layer of meltwater between the ice and substrate. Water-saturated sediment beneath ice can also slide, carrying the ice with it (bed deformation)	Warm-based, temperate glaciers and the margins of some cold-based ice masses
Regelation creep	Occurs when ice melts under pressure, usually because it encounters an obstacle. The meltwater flows around the obstacle and refreezes as the pressure drops	Valley glaciers on steep, rocky slopes at the ice/substrate contact
Internal deformation	Movement between or within individual grains of ice, including grains slipping over each other, melting and recrystallising, and slip within individual grains (intra-granular slip)	The upper parts of ice masses; the only type of motion on cold-based ice masses

Ice velocity is dependent on temperature, as higher temperatures encourage basal sliding. Steep valley glaciers will have higher flow rates than ice sheets on relatively flat topography, and the high-altitude parts of valley glaciers may be cold-based for all or part of the year, slowing velocity. In addition, a glacier experiencing net ablation (negative mass balance) may be rapidly wasting and therefore flowing quickly. Ice velocity also changes along the long profile of a glacier as a result of bed topography (Figure 21). It may also change because of variations in lithology, with easily eroded rocks (sandstone, heavily fractured metamorphic rock) being more prone to bed deformation.

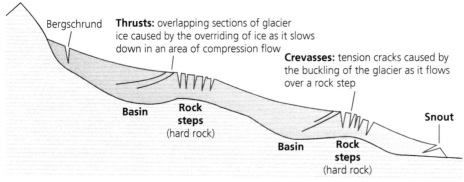

Figure 21 Extensional and compressional ice flow as a result of bed topography

Glacier landform systems

It might seem strange to be studying ice masses when there is none in the UK today. But in geological terms the UK landscape was transformed by ice very recently — the ice only disappeared about 12,500 years ago. Figure 22 shows the variety of glacial and periglacial environments that once existed in Great Britain. These were formed by a combination of four processes:

1 Glacial erosion: the removal of rock material by ice through abrasion and plucking and by meltwater flow from ice margins.

2 Glacial **debris** entrainment: sediment being incorporated into glacial ice and carried along with the ice.

3 Glacial sediment transport: sediment on top of (supraglacial), within (englacial) or at the base (subglacial) of ice being moved.

4 Glacial deposition: when material is released from the ice at the margin or the base of a glacier and simply dumped, or dumped out of meltwater.

Debris refers to any glacially eroded and transported sediment, of any size.

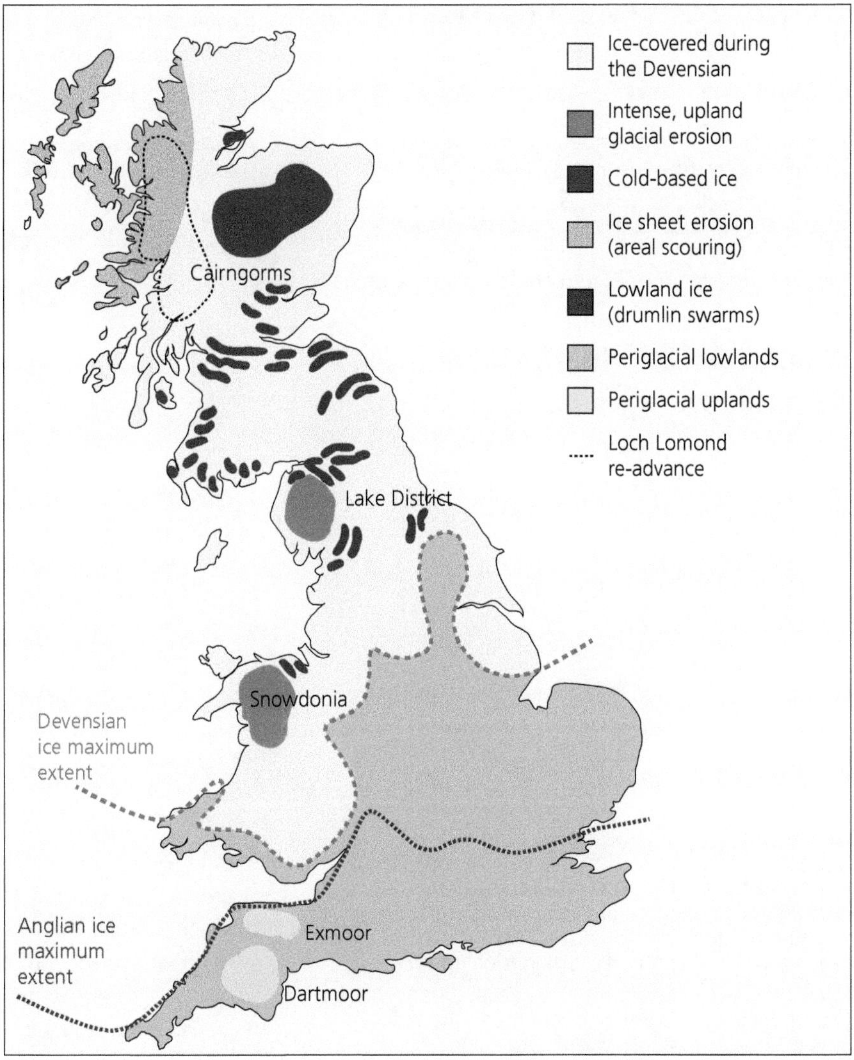

Figure 22 Glacial and periglacial environments of Great Britain

Map legend:
- Ice-covered during the Devensian
- Intense, upland glacial erosion
- Cold-based ice
- Ice sheet erosion (areal scouring)
- Lowland ice (drumlin swarms)
- Periglacial lowlands
- Periglacial uplands
- Loch Lomond re-advance

Map labels: Cairngorms, Lake District, Snowdonia, Devensian ice maximum extent, Anglian ice maximum extent, Exmoor, Dartmoor

These processes occurred in four types of locations, evidence for all of which can be found in the UK landscape.

1 Subglacial areas, which were covered by either warm- or cold-based ice for tens of thousands of years at a time, are dominated by the action of erosion and the erosional landforms it produced.

2 Ice-marginal areas, at the edges of ice sheets and valley glaciers, are areas where deposition from ice was dominant.

3 Proglacial areas, in front of ice masses where meltwater deposition and the action of wind on glacially eroded sediments can be found.

4 Periglacial areas, which had no ice cover but were underlain by permafrost covered in tundra vegetation.

In these locations landforms, which vary in scale, provide evidence of former glacial and periglacial conditions, as shown in Table 22.

Table 22 Glacial landforms by scale

Macro-scale	Ice sheet-eroded knock and lochan landscapes, cirques, arêtes and pyramidal peaks. Glacial troughs, ribbon lakes, till plains, terminal moraines, sandurs
Meso-scale	Crag-and-tail, roches moutonnées, drumlins, kames, eskers and kame terraces, kettle holes
Micro-scale	Features such as striations, glacial grooves and chattermarks, erratics

How do glacial processes contribute to the formation of glacial landforms and landscapes?

- There are a number of glacial erosion processes that produce landforms linked to valley glaciers and large ice masses.
- Depositional processes create landforms at the margins of ice masses and underneath them.
- Meltwater produces distinctive fluvioglacial deposits and landforms.

Glacial erosion

Ice masses are very powerful agents of erosion. As ice flows downslope under the influence of gravity it erodes rock material that is transported to lower levels by ice and then deposited. A rock surface can be eroded by ice in a number of ways (Table 23).

Table 23 Types of glacial erosion

Type	Explanation	Evidence
Abrasion	Rock fragments frozen into the base of moving ice grind down bedrock, eroding it and generating fine eroded material called rock flour	Glacially eroded rock surfaces often have striations: long, parallel scratches 1–10 mm wide, tracing the path of eroding rock fragments
Crushing	The pressure exerted by rock fragments embedded at the base of ice can chip fragments off the bedrock below	Micro-features called chattermarks (wedge- or crescent-shaped depressions) are evidence of fracturing by crushing forces

Exam tip

It is important to know that the UK experienced different types of glacial and periglacial environments at different times in the past.

Knowledge check 26

How long ago did ice cover disappear from the UK?

Erosion is the action of surface processes that remove rock material and begin the process of transporting it to a new location.

Exam tip

There are many different processes and landforms to learn for this section of the specification. Using flash cards will help you to revise them.

Type	Explanation	Evidence
Plucking (quarrying)	Plucking removes chunks of bedrock (in fractured and jointed rock) by freezing water around them so they are pulled away by the ice mass as it moves over the top	Plucking produces steep rock faces with numerous fractures and an angular pattern of 'missing' blocks
Basal melting	This process is similar to abrasion in rivers but differs because water flowing at the base of ice can be under pressure, making abrasion a more powerful force	Deeply eroded channels and potholes can indicate areas where basal meltwater flow occurred in the past

Erosion generates sediment that is incorporated into ice. Two other processes can also be sources of glacial sediment:

1 Freeze–thaw weathering on the rock slopes above valley glaciers is a constant source of angular rock fragments that build up to form scree.

2 **Mass movement** — in the form of rock falls and landslides — is a source of rock material which falls onto the surface of ice and is incorporated into it.

Distinctive landforms of erosion

The variety of ice masses and differences in erosion types produce a range of erosional landforms that reflect their environment of origin (Figure 23). Constrained cirque and valley glaciers erode an impressive landscape of steep, rocky landforms that can be best seen in active areas such as the Alps and relict areas such as the Lake District. Table 24 summarises these landforms.

Table 24 Cirque and valley glacier erosional landforms

Landform	Description	Formation processes
Cirque/corrie	Bowl-shaped depression with a steep, craggy backwall and rock or sediment lip, usually a few hundred metres in diameter	Rotation of an area of accumulated ice causing abrasion and plucking of the bedrock
Arête	A steep rock ridge separating two corries/cirques, formed by their backwalls	Lateral erosion of cirque backwalls gradually narrowing the rock between them
Pyramidal peak	A very steep rock peak in the shape of a pyramid	Formed when three or more cirque back walls and arêtes meet
Glacial trough/ribbon lake	The valley eroded by a glacier, being steep-sided but flat-bottomed (U-shaped). Many are several kilometres long and may today contain a long, thin lake (ribbon lake)	V-shaped river valleys are eroded by ice, and deepening occurs as most erosion is concentrated at the ice base
Truncated spurs	Spurs are projections of bedrock into a river valley, but they are cut (truncated) by glacial erosion to form steep rock faces	Ice shapes former river valleys by erosion cutting into the interlocking spurs of rock and slicing off their ends
Hanging valley	A small glacial valley cutting into a larger one from the side, with its base high above the floor of the larger valley	Ice from a smaller, less deep glacier joins a larger glacier so their bases are at different elevations

Knowledge check 27

Which erosion processes produce chattermarks and striations?

Mass movement is the downslope movement of material under the influence of gravity.

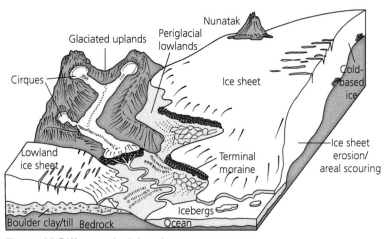

Figure 23 Different glacial environments

Exam tip

Make sure you are clear about the different landforms associated with valley glaciers and ice sheets.

Knowledge check 28

What landform is produced by the lateral erosion of three or more corries?

Ice sheet scouring

Ice sheets are unconstrained ice masses (Figure 23). At their centre they are often cold-based so very little erosion occurs, but towards their edges they are warm-based. Their huge extent and mass mean that they 'sandpaper' vast areas by a process called areal scouring, leading to a relatively low topography landscape.

The west of Scotland, the northwest Highlands and the Isle of Lewis have a very irregular lowland landscape of small, bare-rock, ice-sculpted hills (knocks — similar to roches moutonnées) and small lakes (lochans). This is called 'knock and lochan' topography. This landscape is a result of:

■ resistant bedrock over an extensive area, metamorphic gneiss in the west of Scotland
■ widespread abrasion and plucking by an ice sheet moving in one direction for a long period of time, called areal scouring
■ selective deeper erosion of faults, master joints and weaker areas of rock to form lochans
■ areas of more resistant rock that form the higher knocks.

Knock and lochan landscapes often have a distinct regional alignment of small hills and lakes that follow lines of weakness, e.g. master joints or weaker rock strata. Differential geology can contribute to the formation of two erosional landforms, as shown in Figure 24.

Differential geology refers to the existence of hard and soft rock beds, which ice can erode less/ more easily. It creates landforms that stand above the surrounding topography.

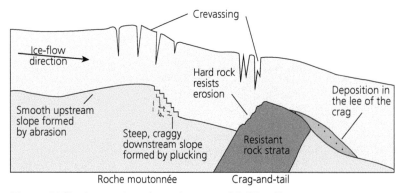

Figure 24 Roche moutonnée and crag-and-tail landforms

Glacial depositional landforms

The most common glacial depositional landforms are moraines. Moraine is
unconsolidated glacial sediment and is unsorted (many different particle sizes) and
angular in shape. Moraines are ice-contact landforms, meaning that the sediment
has been transported only by ice since it was weathered and/or eroded. Four types of
moraine are common:

1 Terminal moraines: these mark the maximum position of the ice and are
   the heap of debris dumped by the snout of a glacier. In the case of valley
   glaciers they are arc-shaped. Ice sheet terminal moraines can be tens of
   metres high and many kilometres long.

2 Recessional moraines: these are similar to terminal moraines but
   marking the positions of an ice mass as it gradually melted and retreated
   during deglaciation.

3 Lateral moraines: rock debris builds up along the edge of valley glaciers
   where the ice meets the valley wall because freeze–thaw weathering
   and mass movement constantly drop debris onto the ice. When the ice
   melts, the debris is dumped in a hummocky line along the valley edge.

4 Medial moraines: when two valley glaciers meet to form one larger glacier,
   their respective lateral moraines merge into a medial moraine running
   down the centre of the larger glacier and when this ice melts, this debris
   is deposited as a line of moraine running down the centre of the glacial trough.

Lowland features

Till is the unconsolidated sediment deposited by a glacier. It is usually a mix of clay,
boulders and gravel:

■ Lodgement till is deposited under the ice and is usually structureless.
■ Ablation till is deposited by melting ice and usually has some evidence of
   deposition in running meltwater.

Under lowland ice sheets (see Figure 23) there is often a layer of glacial sediment that
forms a till plain when the ice melts. The till plain consists of a thick deposit often
called boulder clay and it blankets the bedrock topography below. Much of northern
England is covered in this type of deposit.

Till often forms a landscape of drumlins. A drumlin is a small, rounded hill — shaped
like half an egg ('basket of eggs topography') (Figure 25).

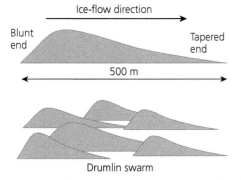

Figure 25 The characteristic shape of drumlins

Unconsolidated means
loose sediment that
has not been cemented
together to form a
sedimentary rock.

Knowledge check 29

Look at Figure 24. What
is the main difference
between roche
moutonnée and crag-and-
tail landforms?

Drumlins often occur together in 'swarms' or 'fields'. Drumlins:

■ are lined up parallel to the direction of ice flow
■ have their 'blunt' end facing into the ice-flow direction and their tapered end facing downstream of the direction of ice flow
■ are typically 100–2000 m long and 50–600 m wide
■ are usually under 50 m high
■ are made of glacially deposited sediments, although some have a rock core.

Drumlins are thought to have formed by deposition under moving ice sheets. As such, they are a subglacial feature. Most are made of lodgement till but some have fluvio-glacial deposits, so water might play a role in their formation. A process involving deformation (moulding and 'streamlining') of lodgement till by relatively fast-moving ice sheets is one way they may form. Moulding by meltwater between ice and till could be responsible for some of the **fluvioglacial sediments** found in some drumlins.

Ice extent and movement

Glacial landforms can be used to reconstruct the extent of ice cover in the past and the direction in which ice masses moved. This is very important for geographers when trying to reconstruct past glacial periods.

One problem is that the landforms created during one glacial period tend to be at best modified, and in many cases destroyed, by later glaciations. Most evidence in the UK is for the Devensian glaciation (110,000 to 12,000 years ago). Evidence for the earlier Anglian glaciation (400,000 to 320,000 years ago) is much rarer. Table 25 summarises the evidence used in ice mass reconstruction.

Table 25 Evidence of former ice mass extent and movement

Ice extent	Ice movement
Terminal and recessional moraines Mark the forward edge of an ice mass so can be used to reconstruct its outer limit, but are vulnerable to subsequent erosion	**Erratics** Very useful for proving that ice flowed from one location to another if the erratic can be linked to its origin
Till plain/boulder clay cover May indicate ice extent, but only for warm-based ice. These deposits can be difficult to date accurately	**Drumlin orientation** The long axes of drumlins are aligned in the direction of ice flow so can be used to show patterns of ice movement
Fluvioglacial deposits Most fluvioglacial action happened at the edge of ice masses so these can indicate areas that were not ice-covered or were at the ice margin	**Crag-and-tail and roche moutonnée** Both landforms have an 'upstream' end so are good indicators of ice flow direction

Meltwater and fluvioglacial landforms

Many glacial depositional landforms are actually the result of the work of ice and water. This is not surprising as all ice eventually melts at the edge of an ice mass and flows away as meltwater. Within ice, meltwater flows in one of three locations, shown in Figure 26. Meltwater not only transports water through the ice mass but also carries sediment with it. This sediment is subject to the attrition processes sediment experiences in rivers so will become more rounded and smaller during its transport by meltwater.

Exam tip

There are many terms for glacial deposits, such as boulder clay, lodgement and ablation till — make sure you are using the correct one.

Fluvioglacial sediments are those that originate through ice erosion but are then transported, eroded and deposited by the meltwater from ice masses.

Knowledge check 30

What is an 'erratic'?

Erratics are rocks and boulders that were glacially eroded in one place but transported by ice and deposited somewhere else. Their rock type does not match the area in which they are deposited.

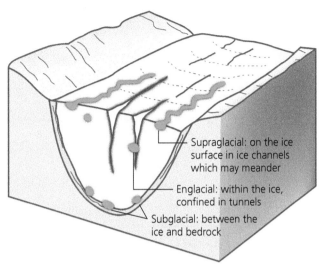

Figure 26 Meltwater movement within a glacier

Glacial and fluvioglacial sediment deposits have very different characteristics because they formed with or without the action of water. These characteristics are summarised in Table 26. Figure 27 (p. 48) compares the characteristics of sediments deposited directly by ice (glacial) and where deposition has involved water (glaciofluvial).

Table 26 Glacial and fluvioglacial sediment characteristics

Characteristic	Explanation	Glacial	Fluvioglacial
Stratification	Stratified sediment has distinct layers within it. Unlayered sediment is described as 'massive', meaning no layers are present	Unstratified, massive	Stratified, representing different phases of deposition
Sorting	Well-sorted sediment has one dominant sediment size, e.g. most is sand-sized particles 1–2 mm in size. Poorly sorted sediments have a wide mix of sediment sizes	Poorly sorted	Well sorted
Imbrication	Imbricated sediments have their long axes (a-axis) aligned in one direction, usually indicating deposition from flowing water	Not imbricated: random axis orientation	Imbricated: dominant axis orientation
Shape	Shape ranges from angular to rounded: water rounds sediment during attrition whereas angular clasts have not been subject to the action of water	Angular	Rounded or subrounded, depending on length of time in transport
Grading	In running water, large sediment is deposited first as the water slows, with fine sediment last. This creates graded sediments, with pebbles, sand, silt and clay being deposited in a sequence from bottom to top	Ungraded	Graded (if a range of clast sizes is present)

All sediment has three **axes**: a, b and c. These correspond to the length (a), width (b) and height (c) of a sediment particle.

Exam tip

You need to be able to recognise sediment characteristics from graphs, rose diagrams and in cross-section.

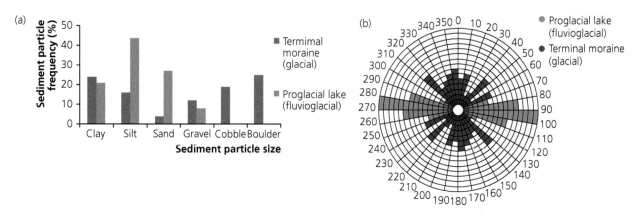

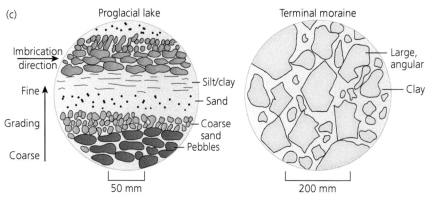

Figure 27 (a) Glacial and fluvioglacial sediments compared; (b) a rose diagram showing the orientation of sediment particle axes (by degrees); (c) cross-sections through proglacial lake and terminal moraine sediments

Fluvioglacial landforms

Water is responsible for landform formation in two situations:

1   Englacial and subglacial meltwater flow helps produce landforms, which are termed ice-contact landforms.

2   Meltwater deposits sediment beyond the ice mass edge, as proglacial landforms.

Proglacial landforms are associated with glacial **outwash plains**.

Table 27 summarises these fluvioglacial landforms.

An **outwash plain** is the area in front of an ice mass where the dominant processes are those of meltwater erosion and deposition. Sandur is another name for a glacial outwash plain.

Table 27 Fluvioglacial landforms

	Landform	Description	Explanation
Ice contact	Kames and kettleholes	Mounds and hillocks of till, with some evidence of meltwater flow; often found next to depressions called kettle holes which may form small ponds and pools	Ice surface sediment is dumped as ice melts beneath it and some reworking by meltwater occurs. Kettle holes are the depressions left as individual ice blocks melt
	Eskers	Long, thin, sinuous gravel ridges (like embankments) often extending hundreds of metres	Formed in ice-walled englacial tunnels (deposited as the ice melts) or constrained subglacial streams
	Kame terrace	Sediment 'benches' on valley sides representing the former ice surface and its contact with the valley wall	Sediment seems to accumulate in ice-bound pools and exhibits stratification and is often well sorted

	Landform	Description	Explanation
Proglacial	Sandur	Extensive, flat sediment plain with numerous meltwater channels draining an ice mass	The dominant processes are river ones, i.e. channel formation and deposition of sediment from river flow, although they are seasonal because of summer ice melt
	Proglacial lake	A lake formed in a depression in front of an ice mass. They are often dammed by moraines	Sediment settles in seasonal layers called varves on the lake bed and exhibits grading
	Meltwater channels	Braided channels and meandering channels running across the surface of a sandur	Formed by meltwater from the ice mass, but more mobile than most river channels because of the lack of stabilising vegetation

How are glaciated landscapes used and managed today?

- Glacial landscapes have high environmental, cultural, economic and scientific value to a range of people.
- There are local and global threats which are damaging fragile active and relict glacial landscapes.
- Managing glacial landscapes to reduce the threats they face is possible but requires action at local, national and international scales.

The value of glaciated landscapes

Many glacial landscapes are relatively isolated areas with low population densities. This includes:

- active high-altitude landscapes such as the Himalayas and the Alps
- active high-latitude landscapes including the periglacial tundra of northern Russia and Canada
- relict upland glacial landscapes such as the Lake District and Cairngorms in the UK.

These areas share some common features, including:

- highly seasonal climates, even in the relict landscapes
- geographical isolation and difficult access
- traditional peoples, often still farming and using resources in a traditional way.

Glaciated landscapes can have strong religious and spiritual associations, giving them a special significance and cultural value to indigenous people (Table 28).

Table 28 Spiritual and religious associations

Himalayan landscape	Arctic Inuit religion
<ul><li>Mount Kailash, which has never been climbed, is sacred to Buddhism, Jainism and Hinduism</li><li>Tibetan prayer flags are often placed on mountains, passes and ridges — showing the importance of the natural world in Buddhism</li><li>In Tibet, pilgrimages take place every year to sacred mountains and lakes</li></ul>	<ul><li>All things, living and non-living have a spirit, a form of animism</li><li>Inuit gods and goddesses often control a particular environmental realm</li><li>Examples include Sedna (goddess of sea animals) and Nanook (god of polar bears)</li><li>Many Inuit religious stories and myths warn against physical dangers in the Arctic</li></ul>

Braided channels are a network of numerous small and large channels that interweave, rather than there being one dominant channel. They are common in proglacial areas because of large variations in meltwater flow and lack of vegetation stabilising channels' banks.

Indigenous people are the original inhabitants of a territory and have strong historical and cultural connections with that area.

Knowledge check 32

What types of landscape features often have religious significance in glaciated landscapes?

> **Synoptic link**
>
> (A): Attitudes of players range from a desire to completely preserve glaciated landscapes for cultural, religious and landscape reasons to those held by people who wish to exploit such landscapes for economic gain. This can lead to conflict.

Glaciated landscapes are important scientifically for a number of reasons:

- Polar regions are ideal locations for studying the upper atmosphere, as the atmosphere is thinner at the poles and less polluted than elsewhere on Earth.
- The dark skies and clear air of polar and high-altitude regions make astronomical research easier than in other places.
- Scientific research has indicated that the Arctic and Antarctic act as 'barometers' for global warming. Their pristine condition and low human population mean that environmental change shows up in these regions first and can be studied.

However, because of globalisation and the increasing availability of transport to once-isolated glacial regions, they are becoming more and more popular with visitors. Arctic and Antarctic cruises, polar trekking, and extreme wilderness sports and recreation are increasingly common in glacial areas that were inaccessible only 30 years ago.

> **Exam tip**
>
> Make sure you know why glaciated landscapes are valuable culturally, scientifically and economically.

Economic activity and importance

Despite low populations, a wide range of economic activity takes place in glaciated landscapes. However, the type of landscape has a major impact on economic activity, as can be seen by a comparison of the economic activities found in the Lake District and Greenland.

Lake District: relict upland glacial

- Farming, especially upland hill sheep farming, which generates low incomes for farmers.
- Tourism is a major income earner, especially in summer months, reflecting the beauty of the landscape and its national park status.
- Forestry is important in some locations because of the Forestry Commission's commercial soft-wood plantations such as Whinlatter.

Greenland: active ice sheet margin

- The largest industry is fishing for cod, prawns and halibut — farming is very limited because of the extreme cold.
- Tourism is of growing importance and includes whale watching and 'iceberg' tourism but it is highly seasonal.
- Mining and oil exploration are growing in importance as technology allows exploitation in harsh environments and global warming makes Greenland more accessible.
- Greenland has built hydroelectric power (HEP) plants in Nuuk, Sisimiut and Ilulissat, providing the towns with electricity.

Biodiversity and natural systems

In many glaciated landscapes, both active and relict, biodiversity is relatively low by global standards. The periglacial Arctic tundra vegetation consists of dwarf shrubs, mosses, lichens, grasses and sedges. As with plant and animal species everywhere, tundra species may contain important genetic material and/or chemical compounds that could benefit food, pharmaceutical or engineering science.

Arctic ice cover and periglacial areas, and the Antarctic, actually play crucial roles in global physical systems despite their apparent isolation. Table 29 explores this.

Biodiversity is the number of plant and animal species in a given area.

Knowledge check 33

The Arctic tundra has a high percentage of the world total of which groups of plant and animal species?

Table 29 The Arctic and global natural systems

The biosphere: animal migration	The climate system: global refrigeration
Migration is an important component of Arctic ecology. Many bird species, including ducks and geese, fly north in the spring to breed in the short Arctic summer and then migrate south in the winter. Caribou undertake similar, but shorter, migrations. Overall species health depends on these migrations	The heat deficit of polar regions is balanced globally by atmospheric circulation and the thermohaline circulation that moves excess tropical heat towards the poles. The high albedo of Arctic sea ice, snow cover and glacier ice reflects 85–90% of incoming solar energy back into space and this has a significant cooling effect on the planet
The water cycle: ice stores	The carbon cycle: sequestration
Water is stored in glacial ice and is gradually released as meltwater, both into the oceans and into rivers. This is an important component of the hydrological cycle. In some locations, such as the Himalayas, summer meltwater from glaciers is a critical component in human water supply	Wetlands, peat lands and lakes cover up to 70% of the Arctic. These ecosystems are an important carbon sink as they store undecomposed organic matter as well as protecting permafrost. Arctic soils, permafrost and sea beds also store methane. Some estimates suggest up to 1400 gigatonnes of methane are stored in the Arctic

Exam tip

Make sure you can link high-latitude glaciated landscapes to wider global systems, including the hydrosphere, atmosphere and biosphere.

Threats to glacial landscapes

Glacial environments are locations of natural hazards but given their low populations, the impacts of natural hazards tend to be small. There are exceptions. One of these is avalanches. Figure 28 (p. 52) shows numbers of deaths from avalanches in the USA since 1950. There is a clear rising trend up to 2010, which can be attributed to:

- increasing numbers of skiers in high-altitude, glacial areas
- the rising popularity of off-piste and extreme skiing
- increased accessibility and improved equipment, encouraging people to venture into isolated areas.

Since 2010, deaths have fallen, perhaps as a result of better management and awareness but perhaps because global warming means fewer skiers on the slopes.

An **avalanche** is a mass movement (landslide) involving snow, ice and rock. Most occur on slopes of 35–45°. The majority are triggered by changes in weather or by people on the slope.

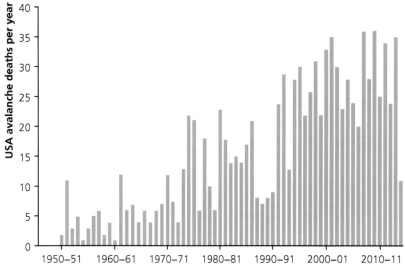

Figure 28 USA avalanche deaths 1950–2019

Glacial outburst floods, or jökulhlaups in Icelandic, are also a threat. These are uncontrolled releases of large volumes of meltwater from underneath ice masses. Often they are triggered by geothermal heating or volcanic eruptions beneath ice sheets and glaciers. While rare, jökulhlaups can cause major damage through flooding and erosion.

■ The 1996 eruption in Iceland of Grímsvötn volcano under the Vatnajökull glacier caused a jökulhlaup with a flood flow rate of 50,000 cubic metres per second, which destroyed parts of Iceland's main ring road.

■ The flood carried ice blocks weighing over 100 tonnes and the initial flood front was 4 m high.

■ In total about 180 million tonnes of sediment were moved by the flood waters.

These floods are fairly common in the Himalaya region, affecting Bhutan, Tibet, Nepal and Pakistan — often as a result of ice-dammed lakes suddenly bursting.

Upland glacial landscapes are subject to a range of human pressures and threats that can degrade their landscape and fragile ecology (Figure 29). Tourism can have a major impact:

■ 120 million tourists visit the Alps each year.

■ Over 300 ski resorts have been developed in the Alps.

■ Extreme sports such as mountain biking, canyoning and paragliding are encroaching on areas previously untouched by tourism.

■ Air pollution levels are rising as vehicle numbers in the Alps increase year on year.

> **Exam tip**
>
> Take care with your spelling, especially when words are of Icelandic origin!

> **Knowledge check 34**
>
> What is a jökulhlaup?

Synoptic link

(F): Degraded landscapes can also be less resilient ones, which are at greater risk of landslides, floods, wildfires and the impacts of climate change in the future.

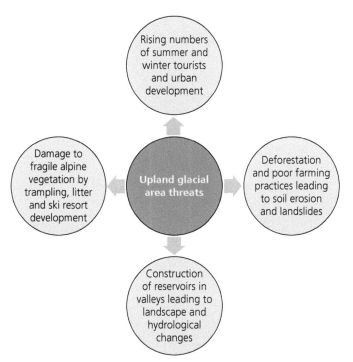

Figure 29 Human activities that threaten upland glacial areas such as the Alps

The narrow but deep glacial troughs in upland glacial landscapes are ideal for dam and reservoir construction. HEP stations in dams can provide power to urban areas and regulate water supply. But dam and reservoir building dramatically alters the landscape and **hydrologic regime** of rivers. The vast Himalaya area is especially vulnerable to this threat, as Table 30 shows.

The **hydrologic regime** of a river refers to its seasonal variations in discharge, both of water and of sediment.

Table 30 Constructed and planned HEP dams in the Himalayas

	Nepal	Pakistan	Bhutan	Indian Himalaya region
Built	15	6	5	74
Under construction	2	7	—	37
Planned	37	35	16	318

Global warming

Rising global temperatures represent the biggest threat to ice masses because:

- higher temperatures lead to negative mass balances and shrinking glaciers, ice caps and ice sheets
- periglacial areas will experience melting permafrost and dramatic changes to the ecosystem and biodiversity of the tundra.

In human terms, the biggest consequence of these changes is likely to be an alteration to the hydrological cycle. In many parts of the world people depend on glacial meltwater for their water supply. In the Peruvian Andes 30% of water comes from this source. As glaciers shrink, so may water supply. However, this may be many decades away because increased melting could temporarily increase supply.

There are other possible consequences:

- increased flood risk if summer ice melt spikes because of extreme heat conditions and causes increases in river discharge
- low meltwater levels limiting HEP dam operation
- low meltwater levels failing to 'flush' rivers clean, leading to higher concentrations of pollution and lower water quality
- changes in the sediment yield of glacial meltwater streams — either becoming choked with excess sediment because of low flow (and increasing the risk of flooding) or dramatically rising during melting and outburst events risking lives and property.

> **Synoptic link**
>
> (A): People rarely seek to degrade active or relict glacial landscapes, however this is occurring indirectly as most humans contribute to global warming and cold environments are among the first to be affected by warming.

River discharge is the volume of water flowing down a river, usually measured in cubic metres per second.

Sediment yield is the volume and type of sediment carried by a river at any one time.

Knowledge check 35

In what way is glacial meltwater important in terms of human quality of life?

Managing threats to glacial landscapes

Threats to glacial landscapes can be managed in a number of ways, depending on the aims and attitudes of those players responsible — the stakeholders (Table 31). The twin aims of conservation and landscape protection versus economic development often conflict, and decisions need to be made in terms of which gets priority. Figure 30 summarises the roles of different stakeholders involved in managing glaciated landscapes.

> **Synoptic link**
>
> (A): Different stakeholders can have radically different views over whether to preserve or exploit glaciated landscapes, and which management approaches (if any) to use.

Table 31 The spectrum of approaches to protection

Protection	Sustainable management	Multiple economic use
Wildlife reserves and wilderness areas designed to limit human activity and maximise landscape and ecological protection	National parks usually try to balance conservation and the needs (economic, social and cultural) of people living in the area	Management may prioritise economic development, especially if it is limited and incomes are low. Conservation is a secondary consideration
The Arctic National Wildlife Refuge (ANWR) in the USA is an example: human activity is limited and oil exploitation is banned. Antarctica, under the Antarctic Treaty, is the ultimate example of this approach. The ANWR is the most biodiverse region in the Arctic	UK national parks such as the Lake District use this method. Economic development is allowed but in a managed, low-key way, and the use of the 'Sandford Principle' ensures that conservation has priority over development if the two aims conflict	Greenland's development of mineral resources, fishing, tourism and HEP (partly using global warming as an economic opportunity) may risk damage to its fragile ice sheet, tundra and ocean environments in a region where economic development opportunities are traditionally very limited

Exam tip

You need to know examples of different types of protection in glaciated landscapes, both active and relict.

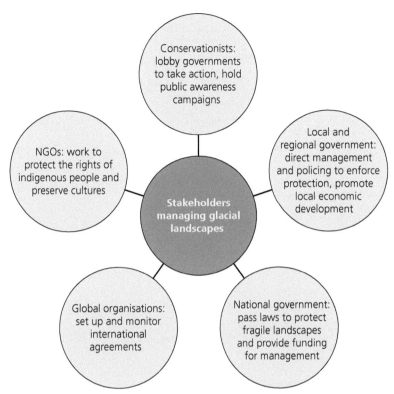

Figure 30 Stakeholders involved in managing glaciated landscapes

Legislative frameworks

A key challenge of managing glacial landscapes is when the landscape requiring protection spans across international borders. This is the case in the Alps, Andes, Himalayas, Arctic and Antarctic. The challenges of this situation are:

■ countries may have contrasting attitudes towards conservation, management and exploitation of resources

■ legal systems may be different between countries

■ financial and human resources available for management may be very different

■ countries may have poor international relations, making agreement hard to reach.

Two examples of **legislative frameworks** are summarised in Table 32 (p. 56). The Antarctic Treaty is a special case. This is because there is no indigenous population in the Antarctic and territory is not owned by any nation (territorial claims have been set aside as part of the Antarctic Treaty).

A **legislative framework** is a set of laws, agreements and treaties (if international) that govern actions by different stakeholders in order to protect and conserve an area.

Table 32 Examples of legislative frameworks

Alpine Convention	Antarctic Treaty
■ Intergovernmental treaty that came into force in 1995 ■ It was signed by the eight Alpine countries — Austria, Germany, France, Italy, Liechtenstein, Monaco, Slovenia and Switzerland — and the EU ■ Attempts to balance development in the Alps with environmental protection ■ Encourages member states to develop policies to manage planning, air pollution, soil conservation, water management, conservation, farming, forests and tourism	■ Covers the entire area south of 60° latitude ■ Came into force in 1961 and has 52 member states ■ Bans military activity in Antarctica ■ Guarantees freedom to use Antarctica for scientific research ■ Limits tourism to day visits only and prevents permanent settlement (other than scientific bases) ■ Fishing is strictly managed and mineral exploitation is banned in Antarctica

Exam tip

Be aware that the Antarctic Treaty is a special case — no other area on Earth is managed in quite the same way.

Co-ordinated approaches

The most significant threat to glacial environments is climate change. The threat can be summarised as:

■ a high risk of 6–8°C warming in the Arctic by 2100, meaning large areas of periglacial permafrost would melt, releasing millions of tonnes of stored carbon

■ 95% of the world's glaciers have a negative mass balance and are retreating, and many will disappear by 2100

■ accelerating melting of the Greenland ice sheet

■ uncertainty about how the Antarctic ice sheet will respond to global warming.

International legislative frameworks and local protection such as national parks can only go so far in protecting glacial environments from the **context risk** of global warming. These actions need to be co-ordinated with a global agreement to reduce greenhouse gas emissions to slow, or stop, global warming:

■ The 1987 Montreal Protocol and 1997 Kyoto Protocol are examples of such actions, but they involved developed countries only.

■ In December 2015 agreement was reached at the COP21 Paris UN climate change conference to reduce emissions. Figure 31 shows what was agreed in Paris. If pledged cuts were implemented, emissions would slow, but not enough to limit global warming to the +2°C many scientists argue is required to avoid major changes to glacial environments.

A **context risk** is one that affects the whole world, such as global warming, so any management actions need to be taken in the context of this wider risk.

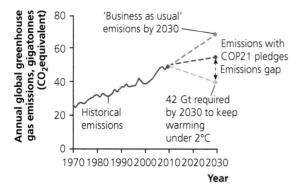

Figure 31 Global greenhouse gas emissions (carbon dioxide equivalent)

Successful management of these unique and fragile landscapes is increasingly challenging, with a need for co-ordinated approaches at global, national and local scales. Further co-ordinated action is likely to be needed to protect the fragile cryosphere.

Synoptic link

(F): There is a high degree of uncertainty over how far global warming will change glacial landscapes, and when. Actions to mitigate climate warming could reduce the threat, but time is running out for sensitive regions such as the Arctic. Without mitigation people and ecosystems face having to rapidly adapt.

Knowledge check 36

Explain why global warming is a context risk.

Summary

- There have been long-term changes to climate and ice cover (glacial and interglacial periods) and shorter-term changes during the Quaternary and Holocene.
- Milankovitch cycles, solar variation and volcanic eruptions are all potential causes of changes to climate and ice cover.
- The cryosphere can be found in high-latitude and high-altitude locations today, but its distribution was different in the past.
- Periglacial environments are widespread in high latitudes today, but also had a different distribution in the past.
- Glacier mass balance is an important concept to understand the balance of accumulation and ablation of an ice mass.
- Ice masses move by different processes and at different rates, which influences the landscapes ice produces.
- Glacial erosion, deposition and fluvioglacial processes all generate specific landforms that together produce contrasting active and relict glacial landscapes.
- Relict and active glacial landscapes are important culturally, economically, and in terms of biodiversity and the water and carbon cycles.
- Human activity, including global warming and more local threats, risks degrading glacial landscapes, which can also experience natural hazards.
- A wide variety of players is involved in managing glaciated landscapes, using a spectrum of approaches with different levels of protection, both locally and using international legislative frameworks.

Coastal landscapes and change

Why are coastal landscapes different and what processes cause these differences?

- The coastal zone, or littoral zone, has a wide variety of distinctive features and landscapes.
- Geological structure is important in generating different coastal landscapes and features.
- Geology, and other factors, are important in affecting rates of coastal retreat.

The coastal zone

Over 1 billion people live on coasts at risk from flooding and about 50% of the world's population live within 200 km of the coast. Many live in the **littoral zone**, which in many cases is a dynamic area of high risk. Risks include coastal flooding and coastal erosion. The littoral zone (Figure 32) can be divided into:

- coast: land adjacent to the sea and often heavily populated and urbanised
- backshore zone: above high tide level and only affected by waves during exceptionally high tides and major storms
- foreshore: where wave processes occur between the high and low tide marks
- nearshore: shallow water areas close to land and used extensively for fishing, coastal trade and leisure
- offshore: the open sea.

The **littoral zone** is the wider coastal zone, including coastal land areas and shallow parts of the sea just offshore.

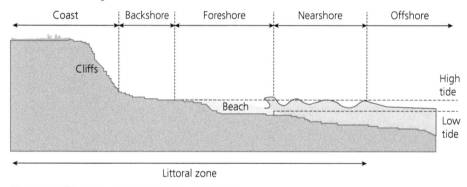

Figure 32 The littoral zone and its subzones

Classifying coastal landscapes

Coasts can be divided into two main types:

1 rocky (or cliffed) coastlines with cliffs varying in height from a few metres to hundreds of metres
2 coastal plains (with no cliffs) where the land gently slopes towards the sea across an area of deposited sediment, often in the form of sand dunes and mud flats.

Cliffs create a sharp distinction between 'land' and 'sea' whereas coastal plains gradually transition from land to sea.

Coasts can be classified according to different physical features and processes (Table 33).

Table 33 Different coastal classifications

Formation processes	Relative sea level change	Tidal range	Wave energy
Primary coasts are dominated by land-based processes, such as deposition at the coast from rivers or new coastal land formed from lava flows	**Emergent coasts** are where the coast is rising relative to sea level, for example as a result of tectonic uplift	Tidal range varies hugely on coastlines, meaning coasts can be: ■ microtidal (tidal range of 0–2 m) ■ mesotidal (tidal range of 2–4 m) ■ macrotidal (tidal range greater than 4 m)	**Low energy** sheltered coasts with limited fetch and low wind speeds resulting in small waves
Secondary coasts are dominated by marine erosion or deposition processes	**Submergent coasts** are being flooded by the sea because of sea level rise and/or subsiding land		**High energy** exposed coasts, facing prevailing winds with long wave fetches resulting in powerful waves

Exam tip

Try to use terms such as 'submergent' and 'emergent' rather than less formal words such as 'sinking' or 'rising'.

Knowledge check 37

Which type of coast is formed by marine erosion and deposition processes?

Rocky coasts

Many coastlines consist of rocky cliffs that vary in height from a few metres (low relief) to hundreds of metres (high relief). High relief cliffs are composed of relatively hard rock. There are two main cliff profile types (Figure 33):

1 Marine erosion dominated: wave action dominates and cliffs tend to be steep, unvegetated and there is little rock debris at the base of the cliff.

2 Subaerial process dominated: not actively eroded at the base by waves; shallower, curved slope and lower relief; surface runoff erosion and mass movement are responsible for the cliff shape.

Subaerial processes include weathering processes (mechanical, chemical and biological), mass movement processes (landslides, rock falls) and surface runoff erosion.

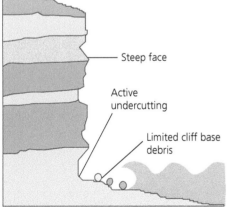

Marine erosion dominated

- Steep face
- Active undercutting
- Limited cliff base debris

Subaerial process dominated

- Curved slope profile
- Lower angle face
- Accumulated debris

Figure 33 Contrasting cliff profiles

Exam tip

Be prepared to sketch cliff profiles and landforms in your answers, as annotated sketches are often quicker than using prose.

Coastal plains

Low-lying, flat (low-relief) areas close to the coast are called coastal plains. Many contain estuary wetlands and marshes, being just above sea level and poorly drained because of the flatness of the landscape.

Coastal plains form when:

- sea level falls, exposing the sea bed of what was once a shallow continental shelf sea, e.g. the Atlantic coastal plain in the USA
- sediment brought from the land by river systems is deposited at the coast causing coastal **accretion** so coastlines gradually move seaward, such as a river delta
- sediment is moved from offshore sources (sand bars) towards the coast by ocean currents.

Accretion refers to the deposition of sediment at a coast that expands the area of land.

Coastal plains are a low-energy environment usually lacking large and powerful waves except on rare occasions such as during hurricanes.

Geological structure

Geological structure is the arrangement of rocks in three dimensions. There are three elements to geological structure:

1 Strata: the different layers of rock exposed in a cliff.
2 Deformation: tilting and folding by tectonic activity.
3 Faulting: major fractures that have moved rocks from their original positions.

Geological structure produces two main types of coast:

1 Concordant or Pacific coasts, when rock strata run parallel to the coastline.
2 Discordant or Atlantic coasts, when different rock strata intersect the coast at an angle, so rock type varies along the coastline.

Discordant coastlines are dominated by headlands and bays. Less resistant rocks are eroded to form bays whereas more resistant geology remains as headlands protruding into the sea. Figure 34 shows the West Cork coast in Ireland.

In this area:

- rock strata meet the coast at 90° in parallel bands
- weak rocks have been eroded, creating elongated, narrow bays
- more resistant rocks form headlands
- especially resistant areas remain as detached islands, such as Clear Island.

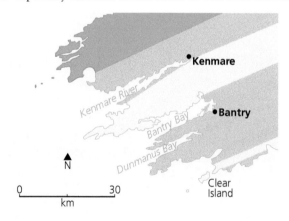

Key

☐ Old red sandstone ▨ Purple mudstone ▨ Limestone
▨ Sandstones and mudstones

Figure 34 The discordant coast of West Cork, Ireland

Headlands and bays change over time because headlands are eroded more than the bays. This happens because of the effect these coasts have on wave crests:

- in deep water wave crests are parallel
- as water shallows towards the coast, waves slow down and wave height increases
- in bays, wave crests curve to fill the bay and wave height decreases
- the straight wave crests refract, becoming curved, spreading out in bays and concentrating on headlands
- the effect of **wave refraction** is to concentrate powerful waves at headlands (so greater erosion) and create lower, diverging wave crests in bays, so reducing erosion.

Concordant coasts are more complex. Different rock strata run parallel to the coast but vary in terms of their resistance to the sea. The most well-known example is Lulworth Cove (Figure 35), where:

- the hard Portland limestone and fairly resistant Purbeck Beds protect much softer rocks landward (the Wealden and Gault beds)
- marine erosion has broken through the resistant beds, and then rapidly eroded a wide cove behind
- there is resistant chalk at the back of these coves, which prevents erosion further inland.

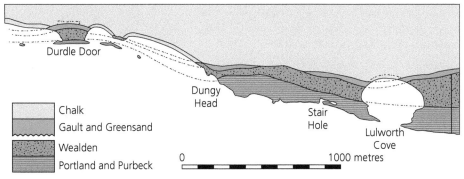

Chalk

Gault and Greensand

Wealden

Portland and Purbeck

0   1000 metres

Figure 35 Lulworth Cove geology

The Dalmatian coast in the Adriatic Sea is another example of a concordant coastline, where:

- the geology is limestone
- the limestone has been folded by tectonic activity into a series of **anticlines** and **synclines**, that trend parallel to the modern coastline.

This underlying structure of upstanding anticlines and lower syncline basins (which would have been eroded by rivers in the past) has been drowned by sea level rise to create a concordant coastline of long, narrow islands arranged in lines offshore. Haff coastlines are also concordant. They are found on the southern edge of the Baltic Sea. Long sediment ridges topped by sand dunes run parallel to the coast just offshore, creating lagoons (the haffs) between the ridges and the shoreline.

Cliff profiles

Cliff profiles are influenced by geology. Two characteristics are dominant:

1 the resistance to erosion of the rock
2 the dip of rock strata in relation to the coastline.

Exam tip

You will need an example of both a discordant and a concordant coast for the exam.

Knowledge check 38

What is the technical name for the different layers of rock found in cliffs?

Wave refraction is the process causing wave crests to become curved as they approach a coastline.

Anticlines and **synclines** are types of geological fold caused by tectonic compression. Anticlines form crests and synclines form troughs.

A **cliff profile** is the height and angle of a cliff face, plus its features such as wave-cut notches or changes in slope angle.

Dip, meaning the angle of rock strata in relation to the horizontal, is important. Dip is a tectonic feature. Sedimentary rocks are formed in horizontal layers but can be tilted by tectonic forces. When this is exposed on a cliffed coastline it has a dramatic effect on cliff profiles, as shown in Figure 36.

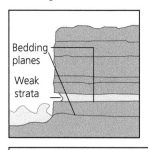 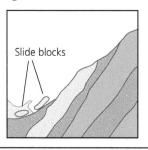 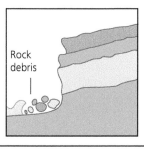 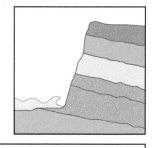

Horizontal dip	Seaward dip high angle	Seaward dip low angle	Landward dip
Vertical or near-vertical profile with notches reflecting strata that are more easily eroded	Sloping, low-angle profile with one rock layer facing the sea; vulnerable to rock slides down the dip slope	Profile may exceed 90°, producing areas of overhanging rock; very vulnerable to rock falls	Steep profiles of 70–80° and producing a very stable cliff with reduced rock falls

Figure 36 Geology and cliff profiles

Other geological features influence cliff profiles and rates of erosion. These include:

- **faults**: either side of a fault line, rocks are often heavily fractured and broken and these weaknesses are exploited by marine erosion
- joints: occur in most rocks, often in regular patterns, dividing rock strata into blocks with a regular shape
- fissures: much smaller cracks in rocks, often only a few centimetres or millimetres long, but they also represent weaknesses that erosion can exploit
- folded rocks: often heavily fissured and jointed, meaning they are more easily eroded.

The location of **micro-features** found within cliffs, such as caves and wave-cut notches, is often controlled by the location of faults and/or strata which have a particularly high density of joints and fissures.

Coastal recession

Coastal recession refers to how fast a coastline is moving inland. This is influenced by many factors, but the key one is lithology or rock type. The three major rock types — igneous, sedimentary and metamorphic — erode at different rates, as shown in Table 34.

Where the rock forming cliffs is **unconsolidated material** (such as sand or boulder clay), rates of recession can be much greater. Boulder clay on the Yorkshire Holderness coast erodes at 2–10 m per year.

Erosion and weathering resistance are influenced by:

- how reactive minerals in the rock are when exposed to chemical weathering
- whether rocks are **clastic** or **crystalline** — the latter are more erosion-resistant
- the degree to which rocks have cracks, fractures and fissures, which are weaknesses exploited by weathering and erosion.

Faults are major fractures in rocks produced by tectonic forces and involve displacement of rocks on either side of the fault line.

Micro-features are small-scale coastal features such as caves and wave-cut notches which form part of a cliff profile.

Unconsolidated material is sediment that has not been cemented to form solid rock, a process known as lithification.

Clastic rocks consist of sediment particles cemented together, **crystalline** rocks are made up of interlocking mineral crystals.

Table 34 Geology and coastal recession rates

Rock type	Examples	Erosion rate and explanation
Igneous	Granite Basalt Dolerite	**Very slow** Less than 0.1 cm per year Igneous rocks are crystalline, and the interlocking crystals make for strong, hard, erosion-resistant rock Igneous rocks such as granite often have few joints, so there are limited weaknesses that erosion can exploit
Metamorphic	Slate Schist Marble	**Slow** 0.1–0.3 cm per year Crystalline metamorphic rocks are resistant to erosion Many metamorphic rocks exhibit a feature called foliation, where all the crystals are orientated in one direction, which produces weaknesses Metamorphic rocks are often folded and heavily fractured, forming weaknesses that erosion can exploit
Sedimentary	Sandstone Limestone Shale	**Moderate to fast** 0.5–10 cm per year Most sedimentary rocks are clastic and erode faster than crystalline igneous and metamorphic rocks The age of sedimentary rocks is important, as geologically young rocks tend to be weaker Rocks with many bedding planes and fractures, such as shale, are often most vulnerable to erosion

Knowledge check 39

Name a type of coastal landform that often forms at the base of a cliff because of a weakness in the rocks, such as a fault line.

Many cliffed coastlines are made of different rock types and so have complex cliff profiles (Figure 37) and experience differential erosion of alternating strata. Cliff profiles can also be influenced by the permeability of strata:

■ Permeable rocks allow water to flow through them. They include many sandstones and limestones.
■ Impermeable rocks do not allow groundwater flow. They include clays, mudstones and most igneous and metamorphic rocks.

Exam tip

Become a bit of a geologist, so you can talk with confidence about different rock types and their properties.

Permeability is important because groundwater flow through rock layers can weaken rocks by removing the cement that binds sediment in the rock together. It can also create high **pore water pressure** within cliffs, which affects their stability. Water emerging from below ground onto a cliff face at a spring can run down the cliff face and cause surface runoff erosion, weakening the cliff.

Pore water pressure is an internal force within cliffs exerted by the mass of groundwater within permeable rocks.

Knowledge check 40

Is sandstone a clastic or crystalline rock?

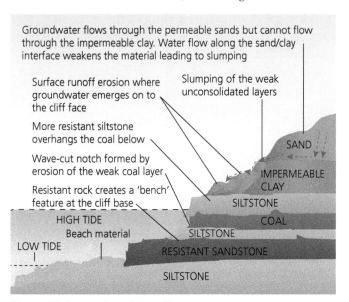

Figure 37 A complex cliff profile

Coastal vegetation

Some coastlines, including coastal sand dunes, salt marshes and mangrove swamps, are protected from erosion by the stabilising influence of plants. Vegetation stabilises coastal sediment in a number of ways:

- Roots bind sediment particles together, making them harder to erode.
- When submerged, plants provide a protective layer so the sediment surface is not directly exposed to moving water and erosion.
- Plants protect sediment from erosion by wind by reducing wind speed at the surface because of friction with the vegetation.

Many plants that grow in coastal environments are specially adapted halophytes (salt tolerant) or xerophytes (drought tolerant).

Plant succession

On a coast where there is a supply of sediment and deposition takes place:

- pioneer plant species will begin to grow in the bare sand or mud
- this forms the first stage of **plant succession**
- each step in plant succession is called a seral stage
- the end result of plant succession is called a climatic climax community.

Coastal climax communities include sand dune ecosystems (psammosere) and salt marsh ecosystems (halosere).

On coastal dunes the succession begins with the colonisation of embryo dunes by pioneer species (Figure 38). Embryo dune pioneer plants:

- stabilise the mobile sand by their root systems
- reduce wind speeds at the sand surface, allowing more sand to be deposited
- add dead organic matter to the sand, beginning the process of soil formation.

Embryo dunes alter the environmental conditions from harsh, salty, mobile sand to an environment that other plants can tolerate. New plant species therefore colonise the embryo dunes, creating a fore dune.

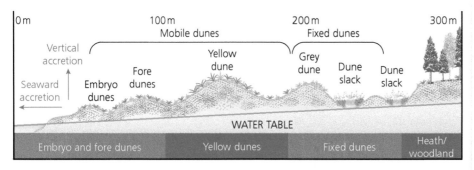

Sea	Bare sand	Sea rocket Saltwort Couch grass Lyme grass	Marram grass	Marram grass Sedge grass	Red fescue Heather Creeping willow	Bramble Pine Birch

Figure 38 Sand dune plant succession

> **Knowledge check 41**
>
> How do plants contribute to preventing erosion of sediment on coasts?

> **Plant succession** means the changing structure of a plant community over time as an area of initially bare sediment is colonised.

A similar process of successional development happens on bare mud deposited in estuaries at the mouths of rivers that are exposed at low tide but submerged at high tide. Estuarine areas are ideal for the development of salt marshes because:

- they are sheltered from strong waves, so sediment (mud and silt) can be deposited
- rivers transport a supply of sediment to the river mouth, which may be added to by sediment flowing into the estuary at high tide.

Exam tip

Notice that the section on vegetation and succession contains language more familiar from biology, which you need to learn to use.

How do characteristic coastal landforms contribute to coastal landscapes?

- Wave types and erosion processes are important in terms of the production of coastal landforms.
- Sediment transport and deposition processes produce coastal landforms, often stabilised by plant succession.
- The sediment cell concept shows how coasts operate as holistic systems.
- Weathering and mass movement subaerial processes are important on some coastlines.

Marine processes and waves

Waves are caused by friction between wind and water transferring energy from the wind into the water. The force of wind blowing on the surface of water generates ripples, which grow into waves when the wind is sustained. In open sea:

- waves are simply energy moving through water
- the water itself only moves up and down, not horizontally
- there is some orbital water particle motion within the wave, but no net forward water particle motion.

Wave size depends on a number of factors:

- the strength of the wind
- the duration the wind blows for
- water depth
- wave fetch.

Waves break as the water depth shallows towards a coastline (Figure 39 on p. 66).

- At a water depth of approximately half the wavelength, the internal orbital motion of water within the wave touches the sea bed.
- This creates friction between the wave and the sea bed, and this slows down the wave.
- As waves approach the shore, wavelength decreases and wave height increases, so waves 'bunch' together.
- The wave crest begins to move forward much faster than the wave trough.
- Eventually the wave crest outruns the trough and the wave topples forward (breaks).

Fetch is the uninterrupted distance across water over which a wind blows and therefore the distance waves have to grow in size.

Exam tip

The process of waves breaking is a classic physical process sequence, which is best explained in logical steps using short sentences.

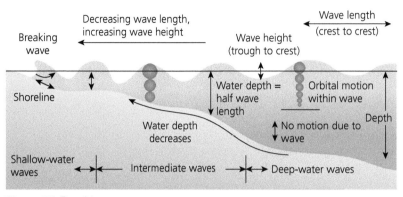

Figure 39 Breaking waves

Knowledge check 42

At what water depth do waves begin to break?

Constructive and destructive waves

Waves breaking on a shoreline do not all have the same shape. There is a basic difference between constructive and destructive waves (Figure 40).

(A) Constructive   **(B)** Destructive

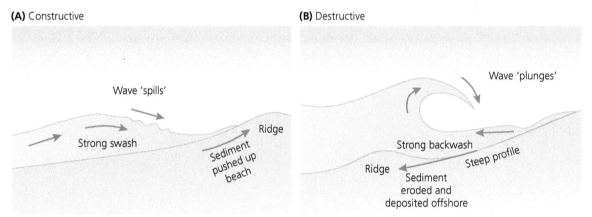

Figure 40 Constructive and destructive waves

Constructive (spilling or surging) waves:

- have a low wave height (less than 1 m) and long wavelength of up to 100 m
- are 'flat' waves with a strong swash but weak backwash
- have strong swash that pushes sediment up the beach, depositing it as a ridge of sediment (berm) at the top of the beach
- have a backwash that drains into the beach sediment.

Destructive (plunging) waves:

- have a wave height of over 1 m and a wavelength of around 20 m
- have strong backwash that erodes beach material and carries it offshore, creating an offshore ridge or berm.

Depending on conditions, beaches experience both constructive and destructive waves over the course of time and this can mean significant changes to beach morphology (beach sediment profile) on different timescales:

- over a day, as a storm passes and destructive waves change to constructive ones as the wind drops

Swash is the flow of water up a beach with a breaking wave. **Backwash** is the water draining down the beach back into the sea. Both can transport sediment.

Beach morphology means the shape of a beach, including its width and slope (the beach profile) and features such as berms, ridges and runnels. It also includes the type of sediment (shingle, sand, mud) found at different locations on the beach.

- between summer and winter (Figure 41)
- when there are changes to climate, e.g. if global warming resulted in the UK climate becoming on average stormier, then destructive waves and 'winter' beach profiles would become more common.

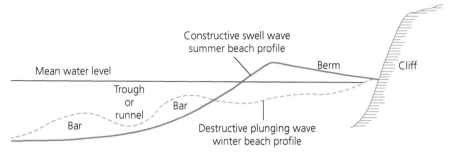

Figure 41 Changes between summer and winter beach profiles

Beaches have landforms that change constantly.

- Storm beaches, high at the back of the beach, result from high energy deposition of very coarse sediment during the most severe storms.
- Berm ridges, typically of shingle/gravel, result from summer swell wave deposition.
- Low channels and runnels between berms.
- Offshore ridges/bars formed by destructive wave erosion and subsequent deposition of sand and shingle offshore.

Marine erosion processes

Waves cause erosion, but erosion is not a constant process. Most erosion occurs during a small number of large storms. There are four main processes that cause erosion (Table 35).

Table 35 Erosion processes

Process name	Explanation
Hydraulic action (wave quarrying)	Air trapped in cracks and fissures is compressed by the force of waves crashing against the cliff face Pressure forces cracks open, meaning more air is trapped and greater force is experienced in the next cycle of compression This process dislodges blocks of rock from the cliff face
Abrasion (corrasion)	Sediment picked up by breaking waves is thrown against the cliff face The sediment acts on the cliff like a tool, chiselling away at the surface and gradually wearing it down
Attrition	As sediment is moved around by waves, the numerous collisions between particles slowly chip fragments off the sediment, making it smaller and more rounded over time
Corrosion (solution)	Carbonate rocks (limestones) are vulnerable to solution by rainwater, spray from the sea and seawater

Erosional coastal landforms

Erosion produces a suite of distinctive coastal landforms. The most well known is the cave–arch–stack–stump sequence (Figure 42, p. 68). The most fundamental process of landform formation on a coastline is the creation of a wave-cut notch. This is eroded at the base of a cliff by hydraulic action and abrasion.

- As the notch becomes deeper, the overhanging rock above becomes unstable and eventually collapses as a rock fall.
- Repeated cycles of notch-cutting and collapse cause cliffs to recede inland.
- The former cliff position is shown by a horizontal rock platform visible at low tide, called a wave-cut platform.

Some landforms are influenced by structural geology. The location of fault lines and major fissures influences the location of caves and therefore arches.

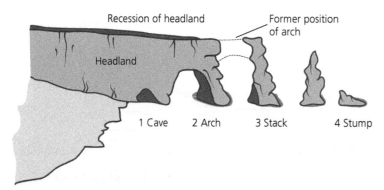

Figure 42 The cave–arch–stack–stump sequence

Sediment transport and deposition

Material eroded from cliffs is transported by the sea as sediment. Transport of sediment happens in four ways (Table 36).

Table 36 Sediment transport processes

Traction	Sediment rolls along, pushed by waves and currents	Pebbles, cobbles, boulders
Saltation	Sediment bounces along, because of either the force of water or of wind	Sand-sized particles
Suspension	Sediment is carried in the water column	Silt and clay particles
Solution	Dissolved material is carried in the water as a solution	Chemical compounds in solution

Waves breaking at 90° to a coast move sediment up and down beaches, but sediment can be transported along the coastline too. The process of sediment transport along the coast is called **longshore drift**. When wave crests approach a coast at an angle (rather than at 90° to the coast), the swash from the breaking waves, and the subsequent backwash, follow different angles up and down the beach in a zig-zag pattern. On most coastlines there is a dominant prevailing wind so over time there is a dominant direction of longshore drift.

Depositional landforms

Longshore drift is a key source of sediment for depositional landforms on coasts. Sediment transported down river systems to the coast or from offshore sources is also important. Sediment is deposited when the force transporting sediment drops. Deposition can occur in two main ways:

Longshore drift is the net transport of sediment along the beach as a result of sediment transport in the swash and backwash.

Knowledge check 44

Name the process that transports sediment along the coastline and contributes to many depositional landforms.

1   Gravity settling occurs when the energy of transporting water becomes too low to move sediment. Large sediment will be deposited first followed by smaller sediment (pebbles → sand → silt).

2   Flocculation is a depositional process that is important for very small particles, such as clay, which are so small they will remain suspended in water. Clay particles clump together through electrical or chemical attraction, and become large enough to sink.

Table 37 summarises the main depositional landforms.

Table 37 Depositional landforms

Landform	Processes
Spit	Sand or shingle beach ridge extending beyond a turn in the coastline, usually greater than 30°. At the turn, the longshore drift current spreads out and loses energy, leading to deposition. The length of a spit is determined by the existence of secondary currents causing erosion, either the flow of a river or wave action which limits its length
Bayhead beach	Waves break at 90° to the shoreline and move sediment into a bay, where a beach forms. Through wave refraction, erosion is concentrated at headlands and the bay is an area of deposition
Tombolo	A sand or shingle bar that links the coastline to an offshore island. Tombolos form as a result of wave refraction around an offshore island, which creates an area of calm water and deposition between the island and the coast. Opposing longshore currents may play a role, in which case the depositional feature is similar to a spit
Barrier beach/bar	A sand or shingle beach connecting two areas of land, with a shallow-water lagoon behind. These features form when a spit grows so long that it extends across a bay, closing it off
Hooked/recurved spit	A spit whose end is curved landward, into a bay or inlet. The seaward (distal) end of the spit naturally curves landward into shallower water and the 'hook' may be made more pronounced by waves from a secondary direction to the prevailing wind
Cuspate foreland	Roughly triangular-shaped features extending out from a shoreline. There is some debate about their formation, but one hypothesis suggests they result from the growth of two spits from opposing longshore drift directions

> **Exam tip**
>
> Most exam questions about coastal landforms will demand an explanation of how they formed, not a description of them.

Vegetation plays a very important role in stabilising depositional landforms. Plant succession, in the form of salt marshes and sand dunes, binds the loose sediment together and encourages further deposition.

The sediment cell model

Coastlines operate as sediment systems consisting of three interlinked components (Figure 43):

1   Sources: places where sediment is generated, such as cliffs or eroding sand dunes. Some sources are offshore bars and river systems and these are an important source of sediment for the coast.

2   Transfer zones: places where sediment is moving alongshore through longshore drift and offshore currents. Beaches and parts of dunes and salt marshes perform this function.

3   Sinks: locations where the dominant process is deposition and depositional landforms are created, including spits and offshore bars.

Sediment cells have inputs, transfers and outputs of sediment. Under natural conditions the systems operate in a state of dynamic equilibrium, with sediment inputs balancing outputs to sinks. For short periods of time — for instance during a major storm that erodes a spit — the system's equilibrium might be disrupted but it will tend to return to balance over time. Negative feedback mechanisms help maintain the balance by pushing the system back towards balance:

- During a major erosion event a large amount of cliff collapse may occur, but the rock debris at the base of the cliff will slow down erosion by protecting the cliff base from wave attack.
- Major erosion of sand dunes could lead to excessive deposition offshore, creating an offshore bar that reduces wave energy, allowing the dunes time to recover.

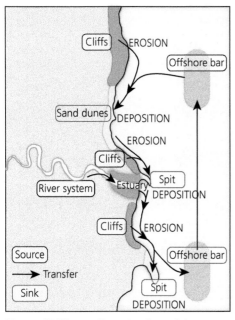

Figure 43 The sediment cell model

Exam tip

The sediment cell concept is important and can be applied to any coast, including those you have studied as case studies.

Knowledge check 45

Why is the sediment cell considered an example of a system?

Subaerial processes

Weathering

Both **weathering** and mass movement are examples of subaerial processes that are important in the development of coastal landforms.

There are three types of weathering important in sediment production:

1 Mechanical weathering breaks down rocks by the exertion of a physical force and does not involve any chemical change.
2 Chemical weathering involves a chemical reaction and the generation of new chemical compounds.
3 Biological weathering often speeds up mechanical or chemical weathering through the action of plants, bacteria or animals.

Weathering is the in situ breakdown of rocks by chemical, mechanical or biological agents. It does not involve any movement.

The **lithology** of coastal rocks is also a major influence, as some rocks are more prone to some types of weathering than others, as outlined in Table 38.

Lithology is the technical term for different rock types.

Table 38 Weathering processes

Type	Process	Explanation	Vulnerable rocks
Mechanical	Freeze–thaw	Water expands by 9% in volume when it freezes, exerting a force within cracks and fissures; repeated cycles force cracks open and loosen rocks	Any rocks with cracks and fissures, especially high on cliffs away from salt spray Freezing is relatively uncommon on UK coasts
	Salt crystallisation	The growth of salt crystals in cracks and pore spaces can exert a breaking force, although less than for freeze–thaw	Porous and fractured rocks, e.g. sandstone The effect is greater in hot, dry climates, promoting the evaporation and the precipitation of salt crystals
Chemical	Carbonation	The slow dissolution of limestone due to rainfall (weak carbonic acid, pH 5.6) producing calcium bicarbonate in solution	Limestone and other carbonate rocks
	Hydrolysis	The breakdown of minerals to form new clay minerals, plus materials in solution, due to the effect of water and dissolved carbon dioxide	Igneous and metamorphic rocks containing feldspar and other silicate minerals
	Oxidation	The addition of oxygen to minerals, especially iron compounds, which produces iron oxides and increases volume, contributing to mechanical breakdown	Sandstones, siltstones and shales often contain iron compounds which can be oxidised
Biological	Plant roots	Trees and plants roots growing in cracks and fissures forcing rocks apart	An important process on vegetated cliff tops which can contribute to rock falls
	Rock boring	There are many species of clams and molluscs that bore into rock, and may also secrete chemicals that dissolve rocks	Sedimentary rocks, especially carbonate rocks (limestone), located in the inter-tidal zone

Knowledge check 46

Name two types of chemical weathering.

Exam tip

The terms 'landslide' and 'mass movement' are umbrella terms for a number of more specific processes.

Weathering contributes to rates of coastal recession in a number of ways. Weathering weakens rocks, making them more vulnerable to erosion or mass movement processes. Some strata may be more vulnerable to weathering than others, contributing to the formation of wave-cut notches and therefore having an effect on overall cliff stability. Rates of weathering are very slow. Even in a hot wet climate, basalt, an igneous rock, weathers at a rate of 1–2 mm every 1000 years.

Mass movement

Weathering and erosion are important on many coasts. On some coastlines **mass movement** is the dominant cause of cliff collapse. There are many different types of mass movement (Table 39), and in a number of types the role of water is very important. Mass movements can be classified in a number of ways, including how rapid the movement is and the type of material (solid rock, debris or soil).

Mass movement refers to the downslope movement of rock and soil. It is an umbrella term for a wide range of specific movements including landslide, rockfall and rotational slide.

Table 39 Mass movement processes

Fall		Rockfalls, or blockfalls, are a rapid form of mass movement
		On coasts, blocks of rock can be dislodged by mechanical weathering, or by hydraulic action erosion
		Undercutting of cliffs by the creation of wave-cut notches can lead to large falls and talus scree slopes at their base
Topple		Geological structure influences topples
		Where rock strata have a very steep seaward dip, undercutting by erosion will quickly lead to instability and blocks of material toppling seaward
Rotational slide/slumping		Mass movements can occur along a curved failure surface
		In the case of a rotational slide, huge masses of material can slowly rotate downslope over periods lasting from days to years
		Water plays an important role in rotational slides
		Rotational slides create a back-scar and terraced cliff profile
Flow		Flows are common in weak rocks such as clay or unconsolidated sands
		These materials can become saturated, lose their cohesion and flow downslope
		Heavy rainfall combined with high waves and tides can contribute to saturation

How do coastal erosion and sea level change alter the physical characteristics of coastlines and increase risks?

- Short- and longer-term sea level changes influence the physical geography of coastlines and increase risks for people.
- Rapid coastal recession happens in some locations because of both physical and human influences.
- Coastal flooding is a significant risk on some coastlines, worsened by global warming but uncertain in terms of magnitude.

Sea level change

Sea level change is complex because both land level (isostatic) and water level (eustatic) can change over time.

■ A rise or fall in water level causes a eustatic change, which is a global change in the volume of sea water in all the world's seas and oceans.
■ Isostatic change is a local rise or fall in land level.

On a coastline, isostatic and eustatic change can happen at the same time. Table 40 summarises the possible changes.

Table 40 Eustatic and isostatic sea level change

Eustatic fall in sea level	Eustatic rise in sea level
During glacial periods, when ice sheets form on land in high latitudes, water evaporated from the sea is locked up on land as ice, leading to a global fall in sea level	At the end of a glacial period, melting ice sheets return water to the sea and sea level rises globally. Global temperature increases cause the volume of ocean water to increase (thermal expansion), leading to sea level rise
Isostatic fall in sea level	**Isostatic rise in sea level**
During the build-up of land-based ice sheets, the colossal weight of ice causes the Earth's crust to sag. When the ice sheets melt, the land surface slowly rebounds upward over thousands of years	Land can 'sink' at the coast because of the deposition of sediment, especially in large river deltas where the weight of sediment deposition leads to very slow 'crustal sag' and delta subsidence

Exam tip

Make sure you use the terms eustatic and isostatic correctly as they are easily confused.

Knowledge check 47

Which leads to a global change in sea level, eustatic or isostatic change?

Since the end of the last ice age 12,000 years ago the UK has felt the impact of continuing sea level change:

■ Scotland is still rebounding upward, in some places by up to 1.5 mm per year, because of **post-glacial adjustment**.
■ In contrast, England and Wales are subsiding at up to 1 mm per year.
■ The UK is 'pivoting', with the south sinking and the north rising.
■ Sea level rise caused by global warming (eustatic) compounds the effect in the south but reduces it in the north.

Emergent coastlines

The extent of isostatic and eustatic changes during and after the last ice age was large:

■ Global sea levels fell by 120 m as ice sheets grew.
■ An equal sea level rise happened over about 1000 years when the ice sheets melted.
■ In North America and northern Europe the post-glacial isostatic adjustment was up to 300 m.

These two linked changes happened at very different rates:

■ Post-glacial sea level rise was very rapid, submerging coastlines.
■ Isostatic adjustment was very slow, with land gradually rising out of the sea.

The effect of these changes was to produce an emergent coast, with landforms reflecting previous sea levels. This is shown in Figure 44 (p. 74).

Post-glacial adjustment (sometimes called post-glacial rebound or post-glacial readjustment) refers to the uplift experienced by land following the removal of the weight of ice sheets.

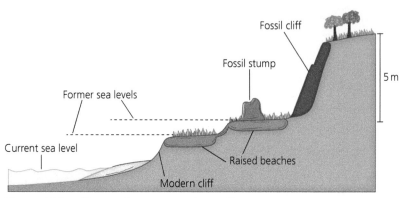

Figure 44 Landforms of an emergent coastline

Submergent coastlines

Coastlines that were never affected by glacial ice cover do not experience post-glacial adjustment. Instead they were submerged (drowned) when post-glacial sea level rose. Submergent coasts are found in southern England and on the east coast of America. The most common coastal landform is a **ria**.

Fjords are submergent landforms found on the coasts of Norway and Canada. Fjords are drowned valleys, but they differ from rias in that:

- the drowned valley is a U-shaped glacial valley
- the fjord is often deeper than the adjacent sea — some are more than 1000 m deep
- at the seaward end of the fjord there is a submerged 'lip', representing the former extent of the glacier that filled the valley.

The east coast of the USA has **barrier island** landforms (Figure 45):

- They may have formed as lines of coastal sand dunes attached to the shore.
- Later sea level rise flooded the land behind the dunes, forming a lagoon, but the dunes themselves were not eroded and so became islands.
- As sea level continued to rise, the dune systems slowly migrated landward.

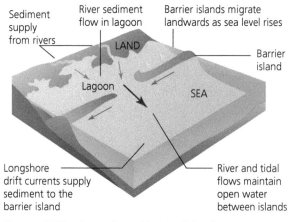

Figure 45 The formation of barrier islands

Exam tip

It is always useful to know located examples of coastlines that are emergent or submergent.

Rias are drowned river valleys in unglaciated areas, caused by sea level rise flooding up the river valley, making it much wider than would be expected based on the river flowing into it.

Barrier islands are offshore sediment bars, usually sand-dune covered, but unlike spits they are not attached to the coast. They are found between 500 m and 30 km offshore and can be tens of kilometres long.

The Dalmatian Coast consists of limestone anticlines and synclines forming a distinctive coastal landscape of parallel hilly islands (anticlines) and elongated bays (synclines). Post-glacial sea level rise submerged this area, creating the elongated bays in places that were once low-lying valleys.

Contemporary sea level change

The rate of sea level rise today is 3.6 mm per year. Figure 46 shows past and projected sea level change.

- Sea level was stable between 1800 and 1870.
- Sea levels rose slowly between 1870 and 1940 but accelerated after that.
- Since 1980, sea level rise has been faster still.
- Between 1870 and today sea level measurements have become more accurate as tide gauges and satellite measurements have become more precise.

Future projections published by the Intergovernmental Panel on Climate Change (**IPCC**) in 2013 range from 28 cm to 98 cm. Some scientists expect global sea level to increase by over 100 cm by 2100.

Knowledge check 48

Name a landform that could be called a 'drowned valley'.

The **IPCC**, part of the United Nations, is a committee of scientists who periodically review the evidence for global warming.

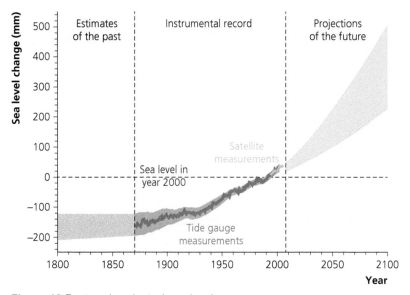

Figure 46 Past and projected sea levels

Sea level is very difficult to predict because there are many factors involved.

- **Thermal expansion** of the oceans depends on how high global temperatures climb.
- Melting of mountain glaciers in the Alps, Himalayas and other mountain ranges will increase ocean water volume.
- Melting of major ice sheets (Greenland, Antarctica) could dramatically increase global sea level.

There is a lot of uncertainty over the contribution of each of these three factors. In addition, sea level can change locally because of tectonic forces. Major earthquakes can force land up or down, sometimes by as much as 2 m.

Exam tip

Learn some data about sea level in the past and projected rises in the future. Make sure you use the correct units, e.g. mm, cm or m.

Thermal expansion, the main driver of sea level rise, occurs because the volume of ocean water increases as global temperatures rise.

Rapid coastal retreat

Rapidly eroding coastlines have the following physical features in common:

- long wave fetch, large destructive ocean waves
- soft geology
- cliffs with structural weaknesses such as seaward rock dip and faults
- cliffs which are vulnerable to mass movement and weathering, as well as marine erosion
- strong longshore drift, so eroded debris is quickly removed, exposing the cliff base to further erosion.

Human actions can make the situation worse and usually involve interfering with the coastal sediment cell (see Figure 43, p. 70). This can happen in a number of ways. The construction of major dams on rivers can trap river sediment behind the dam wall. This then starves the coast of a sediment source, leading to serious consequences.

- The construction of the Aswan High Dam on the River Nile in 1964 reduced sediment volume from 130 million tonnes to about 15 million tonnes per year. Erosion rates jumped from 20–25 metres per year to over 200 metres per year as the delta was starved of sediment.
- The construction of the Akosombo Dam in Ghana in 1965 reduced the flow of sediment down the River Volta from 70 million cubic metres per year to less than 7 million, with major impacts on longshore drift and coastal erosion in Ghana and even in neighbouring countries (Figure 47).

Exam tip

In exam questions, use examples to back up your explanations even if the question is worth only 3–4 marks.

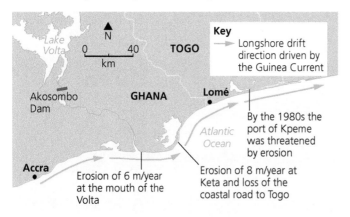

Figure 47 Impacts of the Akosombo Dam on the coastal sediment cell

Dredging is another cause of problems. Sand and gravel are often dredged from the sea bed or rivers for use in construction, i.e. to make concrete or foundations, or to make navigation channels deeper. This removes a sediment source, which can have knock-on effects further along a coast by increasing erosion.

Dredging involves scooping or sucking up sediment from the sea bed or a river bed.

Knowledge check 49

By how much is sea level expected to rise by 2100 as a result of global warming?

Synoptic link

(P): The actions of some players, either upstream or updrift, can have major consequences for others downdrift or where major rivers reach the sea by altering natural systems, which benefits some but creates costs for others.

Variations in erosion rate

Rates of recession are not constant and are influenced by different factors (wind direction and fetch, seasonal changes to weather systems and the occurrence of storms). On the Holderness coast in East Yorkshire, average annual erosion is around 1.25 m, but as Figure 48 shows, there are wide variations in this rate, from 0 m per year to 6 m per year. This is because:

■ coastal defences at Hornsea, Mappleton and Withernsea have stopped erosion
■ these defences have starved places further south of sediment as groynes have interrupted longshore drift
■ erosion rate therefore generally increases from north to south
■ some areas of boulder clay are more vulnerable to erosion than others
■ some cliffs are more susceptible to mass movement.

Erosion of Holderness varies over time.

■ During winter, 2–6 m of erosion is common when storms, combined with spring tides, increase erosion rates.
■ Summer erosion, during periods when constructive waves dominate, is much lower.
■ Northeasterly storms cause most erosion because of the long wave fetch of 1500 km from the north Norwegian coast.

The shape of the beach on Holderness can change and promote erosion. Ords are deep beach hollows parallel to the cliff which concentrate erosion in particular locations by allowing waves to directly attack the cliff with little energy **dissipation**. Ords slowly migrate downdrift by about 500 m per year so the location of most erosion changes over time. Ord locations erode four times faster than locations without ords.

Dissipation is the term used to describe how the energy of waves is decreased by friction with beach material during the wave swash up the beach. A wide beach slows waves down and saps their energy so when they break most energy has gone.

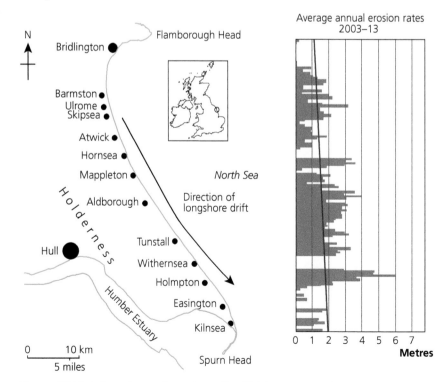

Figure 48 Variations in erosion rate on the Holderness coast

Exam tip

You may have visited a coastline and done fieldwork there. Learn some key facts about its geology and erosion rates.

Coastal flooding risk

Coastal flooding risk is more widespread than rapid erosion risk. Low-lying coastal land the world over is densely populated for several reasons:

- Coastlines are popular with tourists, especially when access to beaches and the sea is easy.
- Deltas and estuaries are ideal locations for trade between up-river places and places along the coast or across the sea.
- Deltas and coastal plains are especially fertile and ideal for farming.

Many of the world's major river deltas, barely a few metres above sea level, are home to some of the world's largest cities (Table 41). Coastal flooding risk in these river delta locations is made worse by a complex set of processes that increase risk (Figure 49).

Table 41 River delta megacities in Asia, 2020

Huang He- Hai	Yangtze	Pearl	Chao Phraya	Ganges-Brahmaputra	Indus
Tianjin (15.6 million)	Shanghai (26.3 million)	Guangzhou (12.1 million)	Bangkok (14.6 million)	Dhaka (21 million)	Karachi (23.5 million)

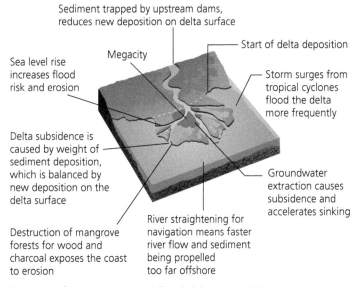

Figure 49 Causes of coastal flood risk in river deltas

In addition to low-lying coastal plains and deltas being at risk from sea level rise, many islands are at risk.

- In the Indian Ocean the Maldives has a population of 340,000 people spread out across 1200 islands. The highest point in the country is 2.3 m above sea level.
- Sea level rise of 50 cm by 2100 would see 77% of the Maldives disappear into the sea.

Storm surges

The most common cause of coastal flooding is a **storm surge** event caused by:

- a depression (low pressure weather system) in the mid-latitudes, e.g. the UK
- a tropical cyclone (hurricane, typhoon) in areas just north and south of the equator.

> **Exam tip**
>
> Good exam answers recognise that some problems, such as flood risk on deltas, have multiple causes rather than one simple cause.

> **Knowledge check 50**
>
> When does most coastal erosion occur?

A **storm surge** is a localised, short-term rise in sea level caused by air pressure change: a 1 millibar fall in air pressure leads to a 1 cm rise in local sea level.

Strong winds from these weather systems push waves onshore, increasing the effective height of the sea. If high tides occur at the same time as the storm surge and onshore winds, sea levels increase even more.

Coastal topography can also have an effect. In the North Sea the coastline narrows into a funnel shape for a storm approaching from the north. The storm surge can be funnelled into an increasingly narrow space between coastlines and as the sea shallows towards the coast, the effect is severe coastal flooding.

The North Sea storm surge of December 2013 was caused by a depression named Xaver. It had major impacts because of winds in excess of 80 mph and a storm surge approaching 6 m.
■ Significant coastal flooding occurred in Boston, Hull, Skegness, Rhyl and Whitby.
■ Scotland's rail network was shut down and 100,000 homes lost power.
■ About 2500 coastal homes and businesses in the UK were flooded.
■ At Hemsby in Norfolk, erosion resulted in several properties collapsing into the sea.
■ Across the countries affected, there were 15 deaths.

Forecasting, warning and evacuation, and improved coastal flood defences limited the scale of the damage.

Bangladesh is especially vulnerable to the impacts of tropical cyclone storm surges for a number of reasons:
■ Much of the country is a low-lying river delta, only 1–3 m above sea level.
■ Incoming storm surges meet out-flowing river discharge from the Ganges and Brahmaputra rivers, meaning river flooding and coastal flooding combine.
■ Intense rainfall from tropical cyclones contributes to flooding.
■ Much of the coastline consists of unconsolidated delta sediment, which is very susceptible to erosion.
■ Deforestation of coastal mangrove forests has removed vegetation that once stabilised coastal swamps and dissipated wave energy during tropical cyclones.
■ The triangular shape of the Bay of Bengal concentrates a cyclone storm surge as it moves north, increasing its height when it makes landfall.

Three major cyclones have struck Bangladesh since 1970 (Table 42). Death tolls have fallen over time because of much improved warnings, the construction of cyclone shelters and better aid response. Vast areas are flooded by these storm surges, forcing millions of people from their homes and farms in the densely populated coastal areas.

Knowledge check 51

What causes a temporary rise in sea level known as a storm surge?

Exam tip

Answers to 12–20-mark exam questions will require more case study detail so make sure you learn the ones you have covered in class or extend the ones in this Student Guide.

Table 42 Bangladesh cyclones

	Storm surge height	Maximum 1-minute sustained winds speed	Lowest air pressure	Deaths and economic losses
1970 Bhola Cyclone	10 m	205 km/h	966 mb	300,000–500,000 US$90 million
1991 cyclone	6 m	250 km/h	918 mb	139,000 US$1.7 billion
2007 Cyclone Sidr	3 m	260 km/h	944 mb	15,000 US$1.7 billion

Climate change and coastal flood risk

The global scientific authority on the link between global warming and coastal flood risk is the IPCC. Its Fifth Assessment Report, published in 2014, came to the conclusions shown in Table 43.

Table 43 Summary of coastal risks from global warming (IPCC, 2014)

Sea level	Delta flooding	Tropical cyclones
Sea level will rise by between 28 cm and 98 cm by 2100, with the most likely rise about 55 cm by 2100	The area of the world's major deltas at risk from coastal flooding is likely to increase by 50%	The frequency of tropical cyclones is likely to remain unchanged, but there could be more of the largest storms
Storm surges	**Wind and waves**	**Coastal erosion**
Storm surges linked to depressions are likely to become more common	There is some evidence of increased wind speeds and large waves	Erosion will generally increase because of the combined effects of changes to weather systems and sea level

- High confidence: the evidence has a high degree of certainty
- Medium confidence: the evidence has some certainty
- Low confidence: the evidence is weak, and uncertainty high

The IPPC report shows that some risks are much more certain than others. The magnitude and timing of these changes are very uncertain.

How can coastlines be managed to meet the needs of all players?

- Erosion, flooding and sea level rise have consequences for people and their property.
- A wide range of management strategies and engineering approaches can be used to reduce risk, but these have advantages and disadvantages.
- Decision making about coastal management can lead to conflict as well as winners and losers.

Consequences of coastal recession for communities

The costs of rapid coastal recession can be classified into three broad categories (Table 44).

Table 44 Costs of coastal recession

Economic costs	Social costs	Environmental costs
Loss of property in the form of homes, businesses and farmland. These are relatively easy to quantify	Costs of relocation and loss of livelihood/jobs (which can be quantified) but also impact on health (such as stress and worry), which is much harder to quantify	Loss of coastal ecosystems and habitats. These are almost impossible to quantify financially but are likely to be small

Losses because of erosion tend to be very localised and costs are specific to those locations. In 2015, farmland in the UK had an average value of about £20,000 per hectare, whereas residential land can vary from £500,000 to £2.5 million per hectare. If roads are lost to erosion they can cost between £150,000 and £250,000 per 100 m to re-route and replace. There are also losses in terms of amenity value and economic losses to businesses if coastlines become unattractive and depopulated.

Synoptic link

(F): The uncertain future of sea levels, and storm risk, associated with global warming creates a problem for coastal communities. Do they try to adapt or mitigate (by building defences)? Both approaches have costs and benefits.

Amenity value is the value in cultural, human wellbeing and economic terms of an attractive environment that people enjoy using.

Economic losses because of erosion are small because:

- erosion happens slowly, with a small number of properties affected over a long period of time
- property that is at risk loses its value long before it is destroyed by erosion because potential buyers recognise the risk
- areas of high-density population, especially towns and villages, tend to be protected by coastal defences.

However, for those communities affected the losses can be significant if, for instance, a whole village is at risk. For coastal people, erosion means:

- falling property values, as the date of eventual loss approaches
- an inability to sell their property because the possibility of loss by erosion is too great
- an inability to insure against the loss (coastal erosion is not covered)
- the loss of their major asset and facing the costs of getting a new home
- an increasingly unattractive environment scarred by collapsing cliffs, failing sea defences and blocked roads and paths.

There is very little help available to people about to lose their homes to the sea. In the UK it consists of 'Coastal Change Pathfinder' projects, which:

- cover the cost of property demolition and site restoration
- provide up to £1000 in relocation expenses such as removal vans and storage
- provide up to £200 in hardship expenses
- have 'rollback' policies, giving people fast-tracked planning approval to build a new home somewhere else.

Consequences of coastal flooding on communities

Coastal floods and storm surges are one-off events that occur perhaps decades apart, whereas erosion is a continual process. Flooding tends to be larger in areal extent and involve greater losses, i.e. in some cases it can be classified as a natural disaster (Table 45).

Table 45 The social and economic impacts of coastal storm surges in developed and developing countries

Example	Cause	Economic costs	Social costs
Netherlands 1953 (North Sea flood)	Mid-latitude depression moving south through the North Sea generating a 5 m storm surge	Almost 10% of Dutch farmland flooded 40,000 buildings damaged and 10,000 destroyed	1800 deaths
UK 2013–2014 winter storms	Coastal and other flooding caused by a succession of depressions and their storm surges	Damage of around £1 billion over the course of the winter	17 deaths (from all causes)
USA 2012 Hurricane Sandy	Landfall of Hurricane Sandy in New Jersey and other US states with a storm surge up to 4 m	US$70 billion in damage 6 million people lost power and 350,000 homes in New Jersey were damaged or destroyed	71 deaths
Philippines 2013 Typhoon Haiyan	One of the most powerful tropical storms ever with a 4–5 m storm surge	Damages of around US$2 billion, centred on the city of Tacloban	At least 6,300 deaths, 30,000 injured

Exam tip

Be careful not to over-exaggerate erosion risk: few people in the UK are threatened by it directly, many more are at risk from coastal flooding.

Environmental refugees

By 2100, in some places sea level rise resulting from global warming will be very difficult to manage. The most at-risk places are low-lying islands including the Maldives, Tuvalu, the Seychelles and Barbados. These small islands have particular risk factors.

- Tuvalu's highest point is 4.5 m above sea level and most land is 1–2 m above sea level.
- 80% of people in the Seychelles live and work at the coast.
- Coral reefs, which act as a natural coastal defence against erosion, are being destroyed by global warming-induced coral bleaching.
- Water supply is limited and at risk from salt-water incursion as sea level rises and groundwater is over-used.
- The islands have small and narrow economies based on tourism and fishing, which are easily disrupted.
- They have high population densities and limited space, so no opportunity for relocation.

The worst-case scenario for Tuvalu, and for part of the Maldives, is that some or all islands will have to be abandoned, creating **environmental refugees**.

Managing coastal recession and flood risk

Hard engineering

The traditional management approach for coastal erosion and/or flooding is to encase the coastline in concrete, stone and steel. This aims to directly stop physical processes altogether (such as erosion or mass movement) or alter them to protect the coast (such as encouraging deposition to build larger beaches). This approach has a number of advantages and disadvantages (Table 46).

Table 46 Advantages and disadvantages of hard engineering

Advantages	Disadvantages
■ Obvious to at-risk people that 'something is being done' to protect them ■ A 'one-off' solution that could protect a stretch of coast for decades	■ Costs are usually very high and there are ongoing maintenance costs ■ Even carefully designed engineering solutions are prone to failure ■ Coastlines are made visually unattractive and the needs of coastal ecosystems are usually overlooked ■ Defences built in one place frequently have adverse effects further along the coast

The economic costs of hard engineering are very high. Groynes cost £150–250 per metre, sea walls £3000–10,000 and rip-rap £1300–6000. Table 47 summarises the most common types of hard engineering.

Synoptic link

(A): Actions by some players can have unforeseen consequences for others, such as the downdrift impacts of groynes which can impose a cost on people who were not expecting it.

Knowledge check 52

Which is more likely to cause widespread economic losses and even deaths, coastal erosion or coastal flooding?

Environmental refugees are communities forced to abandon their homes because of natural processes, including sudden ones such as landslides or gradual ones such as erosion or rising sea levels.

Exam tip

Be careful — it is easy to get bogged down in describing types of coastal defence rather than explaining their purpose or assessing their effectiveness.

Table 47 Hard engineering coastal defences

Type	Construction and materials	Purpose	Impact on physical processes
Rip-rap (rock armour) 	Large igneous or metamorphic rock boulders, weighing several tonnes	Break up and dissipate wave energy Often used at the base of sea walls to protect them from undercutting and scour	Reduced wave energy Sediment deposition between rocks May become vegetated over time
Offshore rock breakwater 	Large igneous or metamorphic rock boulders weighing several tonnes	Forces waves to break offshore rather than at the coast, reducing wave energy and erosive force	Deposition encouraged between breakwater and beach Can interfere with longshore drift
Sea wall 	Concrete with steel reinforcement and deep piled foundations; can have a stepped and/or 'bullnose' profile	A physical barrier against erosion They often also act as flood barriers Modern sea walls are designed to dissipate, not reflect, wave energy	Destruction of the natural cliff face and foreshore environment If reflective, can reduce beach volume
Revetments 	Stone, timber or interlocking concrete sloping structures, which are permeable	To absorb wave energy and reduce swash distance by encouraging infiltration Reduce erosion on dune faces and mud banks	Reduced wave power Can encourage deposition and may become vegetated

➡

Type	Construction and materials	Purpose	Impact on physical processes
Groynes	Vertical stone or timber 'fences' built at 90° to the coast, spaced along the beach	To prevent longshore movement of sediment and encourage deposition, building a wider, higher beach	Deposition and beach accretion Prevention of longshore drift, sediment starvation and increased erosion downdrift

Soft engineering

Soft engineering is an alternative to hard engineering that attempts to work with natural physical systems and processes to reduce the coastal erosion and flood threat. Soft engineering is usually less obvious and intrusive at the coast, and may be cheaper in the long term. However, it is not suitable for all coasts (Table 48).

Knowledge check 53

Which type of hard engineering interferes most with longshore drift?

Table 48 Soft engineering methods

Soft engineering method	Technique	Cost and issues
Beach nourishment	Artificial replenishment of beach sediment to replace sediment lost by erosion, to enlarge the beach so that it dissipates wave energy and reduces wave erosion and increases the amenity value of the beach	Costs of £2 million per km of beach are typical, but ongoing costs are high and sediment must not be sourced from elsewhere in the local sediment cell
Cliff regrading and drainage	Cliff slope angles reduced to increase stability and revegetated to reduce surface erosion. In-cliff drainage reduces pore-water pressure and mass movement risk	Costs of £1 million per 100 m are common. Can be disruptive during construction
Dune stabilisation	Fences are used to reduce wind speeds across the dunes, which are then replanted with marram and lyme grass to stabilise the surface. This reduces erosion by wind and water	Dune fencing costs £400–2000 per 100 m and replanting dunes about £1000 per 100 m. This means that working to maintain natural sand dunes can be very cost-effective in the long term

Sustainable coastal management

Coastal communities around the world face the dynamic nature of the coast's everyday environment. They increasingly face threat from:

- rising global sea levels, but there is uncertainty about the scale and timing of the rise
- increased frequency of storms and the possibility of increased erosion and flooding.

To cope with these threats, communities need to adapt and employ **sustainable coastal management** to ensure the wellbeing of people and the coastal environment (Figure 50).

Sustainable coastal management means managing the wider coastal zone in terms of people and their economic livelihood, social and cultural wellbeing, safety from coastal hazards, as well as minimising environmental and ecological impacts.

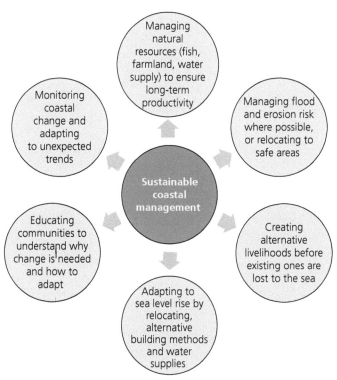

Figure 50 Aspects of sustainable coastal management

Adopting sustainable coastal management may lead to **conflict** because:

- coastal natural resources may have to be used less in order to protect them — so some people lose income
- relocation may be needed where engineering solutions are too costly or not technically feasible
- some erosion and/or flooding will always occur, as engineering schemes cannot protect against all threats
- future trends, such as sea level rise, may change, creating uncertainty and the need to change plan.

Conflict, in the context of coastal management, means disagreement over how the coast should be protected from threats and which areas should be protected. Often conflict exists between different stakeholders, such as residents versus the local council.

Exam tip

Take care when using the word 'conflict'. In a coastal context it means 'disagreement' rather than 'armed conflict'!

Synoptic link

(F): Due to uncertainties in the future, communities may need to both adapt to change and mitigate against it. Either solution on its own may not be affordable in economic terms or for community sustainability.

Integrated Coastal Zone Management (ICZM)

ICZM is a holistic approach increasingly used to manage coasts. It dates from the Rio Earth Summit in 1992 and has a number of key characteristics.

ICZM is coastal management planning over the long term, involving all stakeholders, working with natural processes and using 'adaptive management', i.e. changing plans as threats change.

1 The entire coastal zone is managed, not just the narrow zone where breaking waves cause erosion or flooding. This includes all ecosystems, resources and human activity in the zone.

2 It recognises the importance of the coastal zone to people's livelihoods as, globally, very large numbers of people live and work at the coast — but their activities tend to degrade the coastal environment.

3 It recognises that management of the coast must be sustainable, meaning that economic development has to take place to improve people's quality of life but that this needs to be environmentally appropriate and equitable (benefit everyone).

ICZM works with the concept of littoral cells or sediment cells. The coastline can be divided up into littoral cells and each cell managed as an integrated unit.

■ In England and Wales there are 11 sediment cells.
■ Each cell is managed either as a whole unit or as a sub-unit.
■ In both cases a plan called a Shoreline Management Plan (SMP) is used.
■ The SMP area is further divided into sub-cells.
■ SMPs extend across council boundaries, so many councils must work together on an agreed SMP to manage an extended stretch of coastline.

Policy decisions

In the UK, coastal management is overseen by DEFRA (the government Department for Environment, Food and Rural Affairs). Since DEFRA introduced Shoreline Management Plans in 1995 there have been only four policies available for coastal management. These are very different in terms of their costs and consequences (Table 49).

Table 49 Coastal management policy options

No active intervention No investment in defending against flooding or erosion, whether or not coastal defences have existed previously. The coast is allowed to erode landward and/ or flood	Hold the line Build or maintain coastal defences so that the position of the shoreline remains the same over time
Strategic (managed) realignment Allow the coastline to move naturally (in most cases to recede) but managing the process to direct it in certain areas. Sometimes called 'strategic retreat'	Advance the line Build new coastal defences on the seaward side of the existing coastline. Usually this involves land reclamation

Making decisions about which policy to apply to a particular location is complex. It depends on a number of different factors, including:
■ the economic value of the assets that could be protected, e.g. land
■ the technical feasibility of engineering solutions: it may not be possible to 'hold the line' for mobile depositional features such as spits or very unstable cliffs
■ the cultural and ecological value of land: it may be desirable to protect historic sites and areas of unusual biodiversity
■ pressure from communities: vocal local political campaigning to get an area protected
■ the social value of communities that have existed for centuries.

Littoral cells contain sediment sources, transport paths and sinks. Each littoral cell is isolated from adjacent cells and can be managed as a holistic unit.

Knowledge check 54

What is the name given to a managed unit of coastline that has its own plan, often extending across council boundaries?

Exam tip

Some acronyms, such as ICZM and SMP, are so well known that you can use them in the exam without writing them out in full.

SMPs plan for the future using three time periods, called 'epochs'. These are: up to 2025, 2025–2055 and beyond 2055. A hold-the-line policy applied to an area up to 2025 may become a managed realignment policy after 2025. This is because by 2025 sea level rise is likely to have made 'hold the line' a much more expensive policy to apply.

Cost–benefit analysis

Cost–benefit analysis (CBA) is used to help decide whether defending a coastline from erosion and/or flooding is economically justifiable. An example is Happisburgh in North Norfolk. The policy adopted in this area is 'no active intervention'. This is because to defend the village would have an impact on the wider coastal management plan. Happisburgh would end up as a promontory, blocking longshore drift and causing further erosion downdrift. Longer term, the plan is managed realignment, although this would still involve property being lost to the sea by erosion. Some of the costs and benefits are shown in Table 50.

Table 50 Cost–benefit analysis for Happisburgh

Hold-the-line costs for a 600 m stretch of coastline:
- seawall: £1.8–6 million
- rip-rap: £0.8–3.6 million
- groynes: £0.1–1.5 million

Costs of erosion	Benefits of protection
▪ £160,000 could be available to the Manor Caravan Park to assist in relocating to a new site ▪ Affected residents could get up to £2000 each (a total cost of £40,000–70,000) in relocation expenses plus the cost to the council of finding plots of land to build new houses ▪ Grade 1-listed St Mary's church and Grade 2-listed Manor House would be lost ▪ Social costs as the village is slowly degraded, including health effects and loss of jobs	▪ By 2105, between 20 and 35 houses, with a combined value of £4 million–£7 million (average house price £200,000 in 2015), would be 'saved' from erosion ▪ 45 hectares of farmland, with a value of £945,000, would be saved ▪ The Manor Caravan Park, which employs local people, would be saved

The cost of building coastal defences at Happisburgh is around £6 million. This is very close to the value of property that could be saved, and much higher than the compensation costs payable to local residents. Coastal managers argue that Happisburgh must be seen in the wider context of the whole SMP, further justifying the decision not to defend the village.

Environmental Impact Assessment

Coastal management usually requires an Environmental Impact Assessment (EIA) to be carried out. This is quite separate from any cost–benefit analysis, although it might inform the final CBA. EIA is a process that aims to identify:

- the short-term impacts of construction on the coastal environment
- the long-term impacts of building new sea defences or changing a policy from hold the line to no active intervention or managed realignment.

EIA is wide-ranging and includes assessments of:

- impacts on water movement (hydrology) and sediment flow, which can affect marine ecosystems because of changes in sediment load
- impacts on water quality, which can affect sensitive marine species

Knowledge check 55

Which coastal management policy option involves 'letting nature take its course'?

Exam tip

A worked cost–benefit analysis, such as the one for Happisburgh, could prove useful in the exam as a case study.

Knowledge check 56

What is the main purpose of a cost–benefit analysis on proposed coastal defences?

- possible changes to flora and fauna, including marine plants, fish, shellfish and marine mammals
- wider environmental impacts such as air quality and noise pollution, mainly during construction.

Conflict, winners and losers

Coastal management decisions directly affect people's lives. These effects can be positive or negative, producing perceived:

- winners: people who gain from a decision, either economically (their property is safe), environmentally (habitats are conserved) or socially (communities can remain in place)
- losers: people who are likely to lose property, their business or job, be forced to move, or see the coastline 'concreted over' and view this as an environmental negative.

In some ways this is inevitable because:

- coastal managers produce plans for entire SMP areas, so some areas are protected but others are not
- local councils and government (DEFRA) have limited resources, meaning all places cannot be protected.

There are examples where all stakeholders agree on a course of action. The Blackwater Estuary in Essex is an area of tidal salt marsh and low-lying farmland. Prone to flooding and coastal erosion, the farmland was traditionally protected by flood embankments and revetments. Over the last 30 years it has become clear that responding to rising sea levels and greater erosion by building more and higher coastal defences in places such as Blackwater is not sustainable.

The solution adopted was radical. In 2000 Essex Wildlife Trust purchased Abbotts Hall Farm on the Blackwater Estuary, which was threatened by erosion and flooding. A 4000-hectare managed realignment scheme was implemented by creating five breaches in the sea wall in 2002. This allowed new salt marshes to form inland. The scheme has a number of benefits:

- The Abbotts Hall Farm owners received the market price for their threatened farm.
- The very high costs of a 'hold the line' policy were avoided, but flood risk was reduced.
- Water quality in the estuary improved because of expansion of reed beds that filter and clean the water.
- New paths and waterways were created for leisure activities.
- Additional income streams from ecotourism and wildlife watching were created.
- Important bird (dunlin, redshank, geese) and fish (bass and herring) nurseries were enhanced.

The Blackwater Estuary shows that environmentalists, landowners, coastal managers and local people and businesses can all be kept happy, even when radical plans are adopted.

Synoptic link

(A): Attitudes to coastal management between players depend on a wide range of factors, including how far people prioritise jobs, homes, ecosystems and long-term sustainability. Any particular coastal management strategy is unlikely to please all players.

Coastal management in the developing world

In many parts of the developing world, such as the Maldives, parts of Vietnam and the West African coast, erosion is rapid, often because of a combination of:

- upstream dams reducing sediment supply to the coast and disrupting local sediment cells
- rapid unplanned coastal development, urbanisation and the development of tourist resorts with piecemeal defences and no overall plan
- widespread destruction of mangrove forests for fuelwood and shrimp ponds, exposing soft sediments to rapid erosion.

In many cases, the main 'losers' are the poorest people. Farmers and residents usually lack a formal land title so cannot claim compensation (even if it were available). Coastlines become more vulnerable to sea level rise, the impact of tropical cyclone storm surges and even tsunami. When these disasters strike it is the poorest who lose everything. In many cases it is individual property owners who take responsibility for coastal defences in the absence of local council or government plans.

Summary

- Coastal landscapes and the littoral zone are dynamic, vary in type from rocky to coastal plain and can be high or low energy.
- Geological structure — including concordant and discordant strata, faulting and jointing — is important in shaping the coastal landscape.
- Lithology (igneous, sedimentary, metamorphic, unconsolidated) is a strong influence on rate of coastal recession but vegetation is important in stabilising some coasts.
- Beach profiles vary as a result of constructive and destructive waves, and cliff profiles and landforms are influenced by erosion (abrasion, corrosion, hydraulic action).
- Sediment is transported as part of a sediment cell, which includes landforms produced by deposition and processes such as longshore drift and plant succession.
- Subaerial processes include weathering and mass movement and these are important on some coastlines, especially those with weak geology.
- Sea level change involves a complex interplay of eustatic and isostatic factors and short- and longer-term changes, producing emergent and submergent coastlines with characteristic landforms.
- Some coastlines experience rapid coastal recession resulting from a mix of human and physical factors, while other coasts experience short- and longer-term coastal flood risk.
- Significant local economic losses can result from erosion and/or flooding, with major consequences for vulnerable communities.
- Hard and soft engineering approaches can both be used to manage erosion and flooding, but each has advantages and disadvantages.
- The goal of sustainable coastal management can be achieved through a combination of ICZM, policy decisions, cost–benefit analysis and environmental impact assessment, but the potential for conflict over decisions is high.

■ The water cycle and water insecurity

Water is essential to life on Earth. It is important that we understand its global circulation and distribution, as well as the human demands on it. It is a scarce resource and therefore its use needs to be carefully managed. Failure to do so promises water insecurity.

What are the processes operating within the hydrological cycle from global to local scale?

- The global hydrological cycle is of immense importance to life on Earth.
- The drainage basin is an open subsystem within the global hydrological cycle.
- Water budgets and river systems are strongly influenced by the hydrological cycle.

The global hydrological cycle

A closed system

The global hydrological cycle is the circulation of water around the Earth. It is a closed system of linked processes so there are no external inputs or outputs. For this reason, the amount of global water is finite. The only thing that does change is the state in which the water exists (liquid, vapour or ice). The proportions of global water held in each state vary over time with changes in climate.

The power that drives the global hydrological cycle comes from two sources:

1 solar energy: in the form of heat
2 gravitational energy: causing rivers to flow downhill and precipitation to fall to the ground.

Stores and flows

Figure 51 shows how the global hydrological cycle works. It involves stores, flows and fluxes.

- Stores are 'reservoirs' where water is held. There are four main stores: (1) the oceans, (2) glaciers and ice sheets (cryosphere), (3) surface runoff and (4) the atmosphere. The oceans represent by far the largest store, followed by the cryosphere. Surface runoff consists of rivers and lakes, as well as groundwater. Of these freshwater stores, the cryosphere is the largest, accounting for 69% of all the global freshwater, followed by groundwater (30%). Less than 1% is stored in the biosphere (vegetation and soil moisture).
- Flows are the transfers of water from one store to another. There are four main flows: precipitation, evaporation, transpiration and vapour transport.
- Fluxes are the rates of flow between stores. The greatest fluxes occur over the oceans.

Knowledge check 57

Why is the global hydrological cycle a closed system?

Knowledge check 58

Name a climate change that would alter the proportion of water held in different states.

Exam tip

Sometimes flows are also referred to as transfers.

Surface runoff, in this context, is an umbrella term for a number of land-based stores. These are rivers, lakes, groundwater and the moisture held in soils and vegetation.

Groundwater is the water contained within the soil and underlying rocks, and derived mainly from the percolation of rainwater and meltwater. It is a store, but water also moves through it, hence the term groundwater flow.

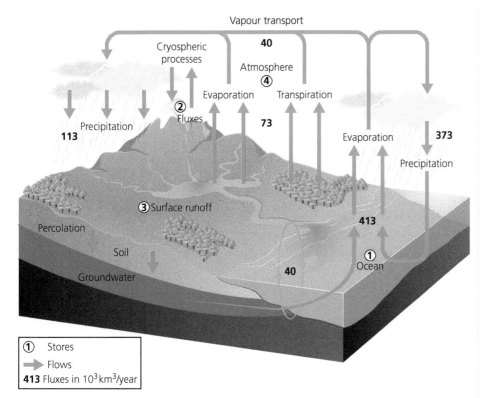

Figure 51 The global hydrological cycle

The global water budget

The global water budget takes into account all the water that is held in the stores and flows of the global hydrological cycle. The most significant feature of the budget is that only 2.5% of it is freshwater; the rest is in the oceans (Figure 52). Even more remarkable is the fact only 1% of all freshwater is 'easily accessible surface freshwater'. Nearly 70% is locked up in glaciers and ice sheets.

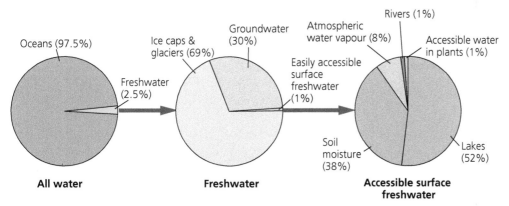

Figure 52 The availability of the world's water

Although water is constantly circulating around the hydrological cycle, each store has a residence time. This is the average time a molecule of water will spend in one of the stores. Residence times vary from 10 days in the atmosphere to 3600 years in the oceans and 15,000 years in an ice cap. Two water stores, **fossil water** and the **cryosphere**, can be thought of as being non-renewable on human timescales.

From a human viewpoint, the most critical feature of the global water budget is that accessible surface freshwater is only 1% of all the world's freshwater (Figure 52). This is the major source of water for human use. The smallness of this figure emphasises the important point that this water is not the abundant resource so many think it is. It is a scarce resource needing careful management.

The drainage basin

An open system

The **drainage basin** is a subsystem within the global hydrological cycle. It is an open system with external inputs and outputs. Since those inputs vary over time, so does the amount of water in the drainage basin (Figure 53). Drainage basins vary in size from that of a small local stream up to a huge river such as the Amazon. The drainage basins of tributary streams and small rivers sit within the drainage basins of larger rivers.

Fossil water is ancient, deep groundwater from pluvial (wetter) periods in the geological past.

The **cryosphere** is made up of those areas of the world where water is frozen into snow or ice.

Knowledge check 59

What are the main stores of easily accessible surface freshwater?

A **drainage basin** is an area of land drained by a river and its tributaries, sometimes referred to as a river catchment. The boundary of a drainage basin is defined by the watershed.

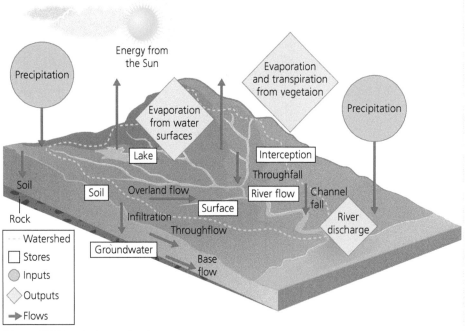

Figure 53 The drainage basin system

Inputs

The main input is precipitation, which can vary in a number of ways. All these characteristics can have a significant impact on the drainage cycle.

- Form: rain, snow or hail. Snow usually has a delayed entry to the drainage system before it melts.
- Amount: this will affect the amount of water in the drainage basin and fluxes within it.

- Intensity: the greater the intensity, the greater the likelihood of flooding.
- Seasonality: this is likely to result in the drainage basin system operating at different flow levels at different times of the year.
- Distribution: this is significant in very large basins, such as the Nile and the Ganges, where tributaries start in different climate zones.

Knowledge check 60

How does solar energy affect the drainage basin system?

Exam tip

There are three types of precipitation: orographic, frontal and convectional. This distinction, made on the basis of causes, is, however, less important than the five bulleted characteristics listed above.

Flows

There are at least seven flows that are important in transferring the precipitation that has fallen on the land into the drainage network (Figure 54):

1. Interception: the retention of water by plants and soils which is subsequently evaporated from or absorbed by the vegetation.
2. Infiltration: the process by which water soaks into or is absorbed by the soil.
3. Percolation: similar to infiltration, but a deeper transfer of water into permeable rocks.
4. Throughflow: the horizontal transfer of water downslope through the soil.
5. Groundwater flow: the very slow transfer of percolated water through permeable and porous rocks. It is related to river base flow.
6. Surface runoff: the movement of water that is unconfined by a channel across the surface of the ground. Also known as overland flow.
7. River or channel flow: takes over as soon as the water enters a river or stream; the flow is confined within a channel.

Exam tip

It is important that you know these seven flows, their distinguishing features and their courses relative to the ground surface. Again, a simple sketch may help.

Knowledge check 61

What is the difference between throughflow and groundwater flow?

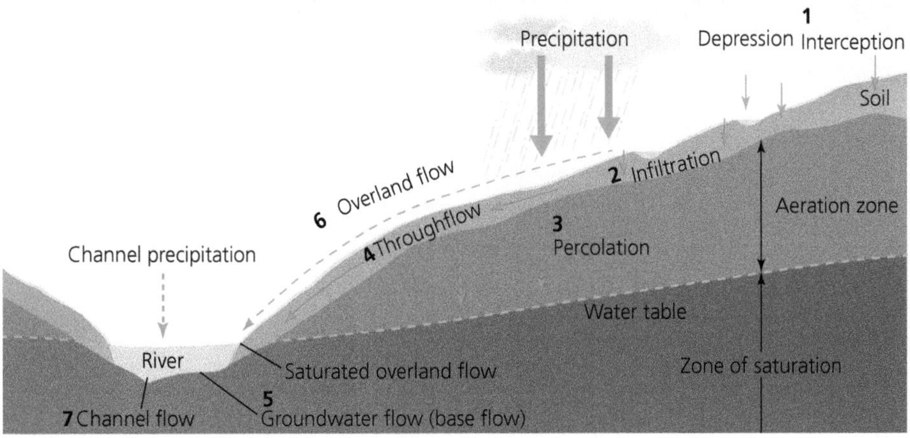

Figure 54 Flows operating within the drainage basin

Outputs

There are three main outputs of the drainage basin.

1 Evaporation: the process by which moisture is lost directly into the atmosphere from water surfaces, soil and rock.

2 Transpiration: the biological process by which water is lost from plants through tiny pores and transferred to the atmosphere.

3 Discharge: (also known as channel flow) the process by which water flows into another, larger drainage basin, a lake or the sea.

Impact of physical factors

Table 51 shows five significant physical factors. Climate mainly impacts the inputs and outputs. The other four factors largely affect the relative importance of the different flows within the system. Of these flows perhaps the most important is surface runoff.

Table 51 Physical factors affecting drainage basin systems

Climate	Climate has a role in influencing the type, amount and seasonality of precipitation and the amount of evaporation, i.e. the major inputs and outputs. Climate also has an impact on the vegetation type
Soils	Soil type determines the amount of infiltration and throughflow and, indirectly, the type of vegetation
Geology	Geology can impact subsurface processes such as percolation and groundwater flow (and, therefore, aquifers). Indirectly, geology affects soil formation
Relief	Relief can impact the amount of precipitation. Slope steepness affects the amount of surface runoff
Vegetation	The presence or absence of vegetation has a major impact on the amount of interception, infiltration and overland flow, as well as on transpiration rates

Impact of human factors

It is mainly human changes to (i) rivers and drainage and (ii) the character of the ground surface (its shape, land use and permeability) that disrupt the drainage basin system, often by accelerating existing processes (Table 52).

Table 52 Some impacts of human activities on drainage basin systems

River management	■ Construction of storage reservoirs holds back river flows ■ Abstraction of water for domestic and industrial use reduces river flows ■ Abstraction of groundwater for irrigation lowers water tables
Deforestation	■ Clearance of trees reduces evapotranspiration but increases infiltration and surface runoff
Changing land use — agriculture	■ Arable to pastoral: compaction of soil by livestock increases overland flow ■ Pastoral to arable: ploughing increases infiltration by loosening and aerating the soil
Changing land use — urbanisation (see more on p. 100)	■ Urban surfaces (tarmac, tiles, concrete) speed surface runoff by reducing percolation and infiltration ■ Drains deliver rainfall more quickly to streams and rivers, increasing chances of flooding

The components of the drainage basin system most affected by humans are:

- evaporation and **evapotranspiration**
- interception
- infiltration
- groundwater
- surface runoff.

Amazonia

The Amazon basin contains the world's largest area of tropical rainforest. Deforestation here has disrupted the drainage basin cycle in a number of ways, including:

- lowering humidity, so fewer clouds form and precipitation decreases
- more surface runoff and less infiltration
- lower transpiration but more rapid evaporation
- more soil erosion and sediment being transported into the rivers.

Knowledge check 62

What would be the effects of afforestation on drainage basin flows?

Water budgets and river systems

Water budgets

A water budget is the annual balance between precipitation, evapotranspiration and runoff. It is calculated from the formula:

$$P = E + R \pm S$$

where P is precipitation, E is evapotranspiration, R is runoff and S represents changes in storage over a period of time, usually one year.

The balance can be calculated at various scales, from global to local. Water budgets at a national or regional scale provide a useful indication of the amount of water that is available for human use (for agriculture, domestic consumption, etc.). At a local scale, water budgets can inform about **available soil water**. This is valuable to users, such as farmers, who can use it to identify when irrigation might be required, and how much.

Of course, as the caption for Figure 55 (p. 96) implies, soil water availability varies considerably from one climatic region to another.

Evapotranspiration is the combined effect of evaporation and transpiration, such as occurs from most vegetated surfaces.

Available soil water is the amount of water that can be stored in the soil and is available for growing crops.

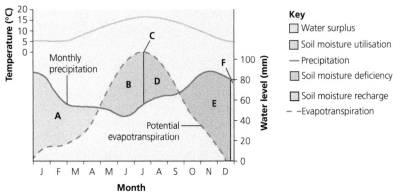

A Precipitation > potential evapotranspiration. Soil water store is full and there is soil moisture surplus for plant use. Runoff and groundwater recharge.

B Potential evapotranspiration > precipitation. Water store is being used up by plants or lost by evpaoration (soil moisture utilisation).

C Soil moisture store is now used up. Any precipitation is likely to be absorbed by the soil rather than produce runoff. River levels fall or rivers dry up completely.

D There is a deficiency of soil water as the store is used up and potential evapo-transpiration > precipitation. Plants must adapt to survive, crops must be irrigated.

E Precipitation > potential evapotranspiration. Soil water store starts to fill again (soil moisture recharge).

F Soil water store is full, field capacity has been reached. Additional rainfall will percolate down to the water table and groundwater stores will be recharged.

Figure 55 A water budget graph for a cool temperate location

River regimes

A river regime is the annual variation in the discharge of a river at a particular point and is measured in cumecs (cubic metres per second). River regimes are influenced by a number of factors (Figure 56):

- the amount, seasonality and intensity of precipitation
- the temperatures, influencing the timing of spring snow meltwater and rates of evaporation in summer
- the geology and soils, particularly their permeability and porosity; groundwater in permeable rocks is gradually released into the river as base flow
- the type of vegetation cover: wetlands can hold water and release it slowly into the river
- human activities aimed at regulating a river's discharge.

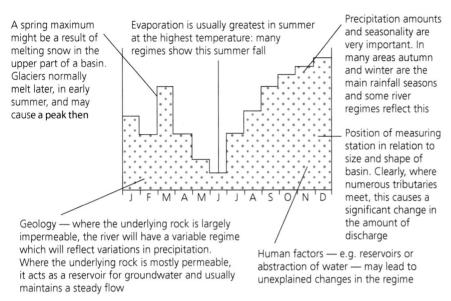

A spring maximum might be a result of melting snow in the upper part of a basin. Glaciers normally melt later, in early summer, and may cause **a peak then**

Evaporation is usually greatest in summer at the highest temperature: many regimes show this summer fall

Precipitation amounts and seasonality are very important. In many areas autumn and winter are the main rainfall seasons and some river regimes reflect this

Position of measuring station in relation to size and shape of basin. Clearly, where numerous tributaries meet, this causes a significant change in the amount of discharge

Geology — where the underlying rock is largely impermeable, the river will have a variable regime which will reflect variations in precipitation. Where the underlying rock is mostly permeable, it acts as a reservoir for groundwater and usually maintains a steady flow

Human factors — e.g. reservoirs or abstraction of water — may lead to unexplained changes in the regime

Figure 56 Factors affecting a river's regime

Figure 57 underlines the importance of climate in determining the basic character of river regimes.

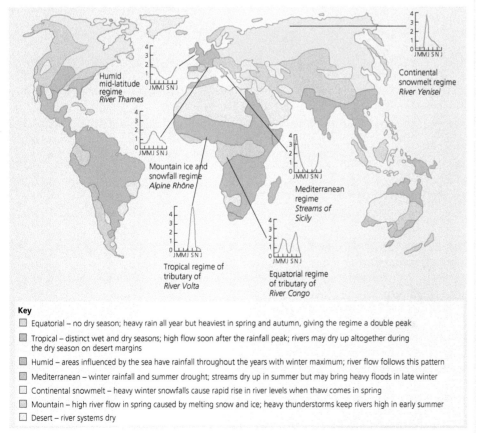

Humid mid-latitude regime
River Thames

Continental snowmelt regime
River Yenisei

Mountain ice and snowfall regime
Alpine Rhône

Mediterranean regime
Streams of Sicily

Tropical regime of tributary of
River Volta

Equatorial regime of tributary of
River Congo

Key

☐ Equatorial – no dry season; heavy rain all year but heaviest in spring and autumn, giving the regime a double peak

▨ Tropical – distinct wet and dry seasons; high flow soon after the rainfall peak; rivers may dry up altogether during the dry season on desert margins

▨ Humid – areas influenced by the sea have rainfall throughout the years with winter maximum; river flow follows this pattern

▨ Mediterranean – winter rainfall and summer drought; streams dry up in summer but may bring heavy floods in late winter

☐ Continental snowmelt – heavy winter snowfalls cause rapid rise in river levels when thaw comes in spring

▨ Mountain – high river flow in spring caused by melting snow and ice; heavy thunderstorms keep rivers high in early summer

☐ Desert – river systems dry

Figure 57 River regimes and climatic regions

Knowledge check 64

Study Figure 57 and compare the river regimes of the River Yenisei and the River Rhône. Suggest reasons for the differences.

Storm hydrographs

Whereas river regimes are usually graphed over the period of a year, storm hydrographs show discharge changes over a short period of time, often no more than a few days. The storm hydrograph plots two things: the occurrence of a short period of rain (such as a storm) over a drainage basin and the impact of this on the discharge of the river.

Figure 58 shows the main features of a storm hydrograph:
- Once the rainfall starts, river discharge begins to rise; this is known as the rising limb.
- Peak discharge is reached some time after the peak rainfall because the water takes time to move over and through the ground to reach the river.
- The time interval between peak rainfall and peak discharge is known as the lag time.
- The falling or recessional limb occurs after the peak, as discharge returns to normal.
- Eventually the river's discharge returns to its normal level or base flow.

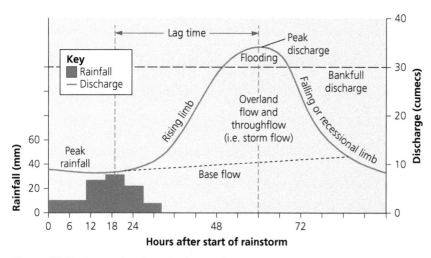

Figure 58 Features of a storm hydrograph

The shape of a storm hydrograph of the same river may vary from one rain event to the next. This variation is closely linked to the nature of the rainfall event. The shape of the hydrograph also varies from one river to another. This is the result of the physical characteristics of individual drainage basins (Table 53).

Some hydrographs have very steep limbs, especially rising limbs, a high peak discharge and a short time lag. These are often referred to as 'flashy' hydrographs. In contrast, there are some hydrographs with gently inclined limbs, a low peak discharge and a long lag time. These are often called 'delayed' or 'flat' hydrographs.

Table 53 Interacting factors affecting the shapes of two different storm hydrographs

Factor	'Flashy' river	'Flat' river
Description of hydrograph	Short lag time, high peak, steep rising limb	Long lag time, low peak, gently sloping rising limb
Weather/climate	Intense storm that exceeds the infiltration capacity of the soil Rapid snowmelt as temperatures suddenly rise above zero Low evaporation rates due to low temperatures	Steady rainfall that is less than the infiltration capacity of the soil Slow snowmelt as temperatures gradually rise above zero High evaporation rates due to high temperatures
Rock type	Impermeable rocks, such as granite, which restrict percolation and encourage rapid surface runoff	Permeable rocks such as limestone, which allow percolation and so limit rapid surface runoff
Soils	Low infiltration rate, such as clay soils (0–4 mm/h)	High infiltration rate, such as sandy soils (3–12 mm/h)
Relief	High, steep slopes that promote surface runoff	Low, gentle slopes that allow infiltration and percolation
Basin size	Small basins tend to have more flashy hydrographs	Larger basins have more delayed hydrographs; it takes time for water to reach gauging stations
Shape	Circular basins have shorter lag times	Elongated basins tend to have delayed or flatter hydrographs
Drainage density	High drainage density means more streams and rivers per unit area, so water will move quickly to the measuring point	Low drainage density means few streams and rivers per unit area, so water is more likely to enter the ground and move slowly through the basin
Vegetation	Bare/low density, deciduous in winter, means low levels of interception and more rapid movement through the system	Dense, deciduous in summer, means high levels of interception and a slower passage through the system; more water lost to evaporation from vegetation surfaces
Pre-existing (antecedent) conditions	Basin already saturated from previous rain, water table high, soil saturated so low infiltration/percolation	Basin dry, low water table, unsaturated soils, so high infiltration/percolation
Human activity	Urbanisation producing impermeable concrete and tarmac surfaces Deforestation reduces interception Arable land, downslope ploughing	Low population density, few artificial impermeable surfaces Reforestation increases interception Pastoral, moorland and forested land

Exam tip

Draw a mind map of the factors affecting the shape of a storm hydrograph.

Knowledge check 65

Apart from steady rainfall, what other factors contribute to a 'delayed' hydrograph?

Urbanisation

When it comes to evaluating the factors affecting the character of storm hydrographs, particularly their 'flashiness', none is more important than urbanisation. This is because it changes the character of the land surface. Its effects on hydrological processes include:

■ Construction work leads to the removal of the vegetation cover. This exposes the soil and increases overland flow.

■ Bare soil is eventually replaced by a covering of concrete and tarmac, both of which are impermeable and increase surface runoff.

■ The high density of buildings means that rain falls on roofs and is then quickly fed into drains by gutters and pipes.

■ Drains and sewers reduce the distance and time rainwater travels before reaching a stream or river channel.

■ Urban rivers are often channelised with embankments to guard against flooding. When floods occur, they can be more devastating.

■ Bridges can trap floodwater debris and act as dams, prompting upstream floods.

Exam tip

Draw a mind map of the ways in which urbanisation affects hydrological processes.

The overall impact of urbanisation is to increase flood risk. The problem is made worse by the fact that many towns and cities are located close to rivers. Historically, this was for reasons of water supply and sewage disposal. Often the original town centre was located at a point where a river could be easily crossed.

Planners have become important players in managing the impacts of urbanisation on flood risk. This is because:

■ many towns and cities are naturally prone to flooding because of their locations
■ of the numbers of people who live in urban places and who therefore need protection
■ of the huge amount of money invested in urban property.

Flood risk management involves such actions as:
■ strengthening the embankments of streams and rivers
■ putting in place flood emergency procedures
■ steering urban development away from high-risk areas such as floodplains.

Synoptic link

(P): Town planners are important players in managing flood risk, but they have to manage conflicting demands from property owners, environmental groups and taxpayers.

Knowledge check 66

Identify some other players involved in the issue of urban flooding.

What factors influence the hydrological system over short- and long-term timescales?

- Short-term deficits within the hydrological cycle (i.e. droughts) are the result of physical processes, but they can have significant impacts on people.
- Short-term surpluses within the hydrological cycle (i.e. floods) are the outcome of physical processes and can have significant impacts on people.
- In the longer term, climate change can have a significant impact on the hydrological cycle, both globally and locally.

Deficits within the hydrological system

Drought is defined in meteorological terms as a shortfall or deficiency of water over an extended period, usually at least a season. Meteorological drought is sometimes distinguished from hydrological drought. The latter is characterised by reduced river flow, lowered groundwater levels and reduced water stores.

Drought can and does hit agricultural productivity particularly hard. This, in turn, can lead quickly to food shortages, famine and starvation in vulnerable developing countries.

While the causes of drought are basically physical, human activities often worsen the impacts of drought.

The physical causes of drought

The physical causes of drought are complex. Fundamentally drought is caused by a period of below average precipitation, but there are several possible triggers for this. These include short-term changes to weather patterns, periodic climate cycles such as the ENSO (see below) and longer-term climate trends driven by global warming.

Research suggests that **sea surface temperature anomalies** are an important causal factor in short-term precipitation deficits.

El Niño–Southern Oscillation (ENSO)

Temperature anomalies provide the key to the El Niño–Southern Oscillation, which in turn triggers drought. Figure 59 (p. 102) shows normal conditions in the Pacific basin and then conditions during an El Niño event. When this happens, cool water normally found along the coast of Peru is replaced by warmer water.

At the same time, the area of warmer water further west, near Australia and Indonesia, is replaced by cooler water. El Niño events periodically occur every 3–7 years and usually last for about 18 months. El Niño events seem to trigger very dry conditions in several global locations, usually in the second year. For example, the monsoon rains in India and Southeast Asia often fail.

Sea surface temperature anomalies relate to how much temperatures of the sea surface, recorded at a particular time, differ from the long-term average. Anomalies may be positive or negative. A positive anomaly occurs when the observed temperature is warmer than the average. A negative anomaly is when the observed temperature is cooler than the average.

A normal year

Warm, moist air rises, cools and condenses, forming rain clouds

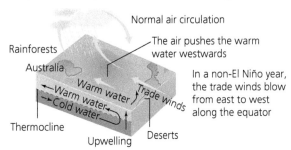

An El Niño year

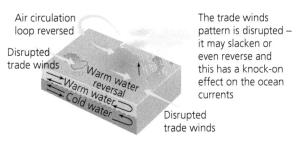

A La Niña year

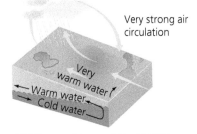

Figure 59 The workings of the ENSO

La Niña episodes may, but not always, follow an El Niño event. They involve the build-up of cooler-than-usual subsurface water in the tropical part of the Pacific. This situation can also lead to severe drought conditions, particularly on the western coast of South America.

Human activity and the drought risk

Desertification in the Sahel

The point has already been made, and is worth repeating, that people are not the cause of drought, but their actions can make droughts worse and more severe. This is well illustrated by **desertification** in the Sahel region of Africa, stretching from Mauritania eastwards to Ethiopia (Figure 60).

Knowledge check 67

Which part of the globe is most affected by (a) El Niño events and (b) La Niña episodes?

Desertification is the process by which once-productive land gradually changes into a desert-like landscape. It usually takes place in semi-arid land on the edges of existing deserts. The process is not necessarily irreversible.

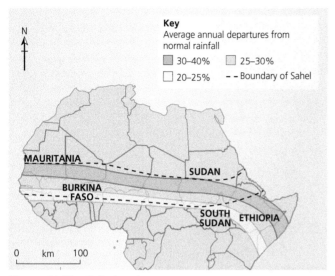

Figure 60 The Sahel region

The causes of desertification are essentially natural. They set in motion a downward spiral that runs roughly as follows:

- Rainfall patterns change, with rainfall becoming less reliable, seasonally and annually (note the annual anomalies in Figure 60). The occasional drought year sometimes extends to several years.
- The vegetation cover becomes stressed and begins to die, leaving bare soil.
- The bare soil is eroded by wind and an occasional intense shower.
- When rain does fall, it is often only for short, intense periods. This makes it difficult for the remaining soil to capture and store it.

Human factors do not cause drought but they act like a feedback loop. Humans enhance the impacts of droughts by the over-abstraction of surface water from rivers and ponds, and of groundwater from aquifers. Key human factors encouraging this are:

- population growth: rapid population growth puts pressure on the land to grow more food; migration from disaster zones (war/conflict) can add to this pressure
- overgrazing: too many goats, sheep and cattle destroy the vegetation cover
- overcultivation: intense use of marginal land exhausts the soil and crops will not grow
- deforestation: trees are cut down for fuel, fencing and housing. The roots no longer bind the soil and erosion ensues.

In the case of the Sahel, the situation has been made worse by frequent civil wars. Crops, livestock and homes have been deliberately destroyed.

Drought in Australia

Drought is a recurrent annual feature in Australia, with up to 30% of the country affected by serious or severe rainfall deficiency:

- El Niño events are strongly linked to the occurrence of drought.
- Droughts are becoming more frequent and more severe.

Exam tip

The Sahel is the classic example of desertification. Be sure that you are fully aware of the human contribution to the process.

Knowledge check 68

Of the four human factors contributing to desertification, which do you think is the most significant?

- To date, the worst event by far has been the 'Big Dry' arguably lasting from 1996 to 2011, but recurring in 2016–2019. This has been assessed as a 1-in-1000-year event.
- ENSO events 1997, 2001 and 2006 contributed to the Big Dry (or Millennium Drought), but there is growing recognition that longer-term climate change driven by global warming may be increasingly important.
- Unlike the Sahel, Australia has not followed the same downward spiral of desertification. Careful management of scarce water resources, and sorting out the competing demands of irrigation and urban dwellers, has stopped this happening. Other actions include the large-scale recycling of grey water, constructing desalinisation plants and devising new water conservation strategies.

The ecological impacts of drought

The concept of ecological resilience is crucial here. Here the focus is on the impact of droughts on the resilience of two ecosystems — wetlands and forests.

Wetlands

Wetlands cover about 10% of the Earth's land surface and until 50 years ago they were considered as wastelands, only good for draining and infilling to provide building land. However, it is now understood that wetlands perform a number of important functions: from acting as temporary water stores to the recharging of aquifers, from giant filters trapping pollutants to providing nurseries for fish and feeding areas for migrating birds.

Drought can have a major impact on wetlands. With less precipitation there will be less interception (as vegetation becomes stressed), as well as less infiltration and percolation. Water tables will fall. Evaporation will also increase. This, together with the decrease in transpiration, will reduce the valuable functions performed by wetlands.

While droughts pose a threat to wetlands, the major challenge to their survival remains artificial drainage.

Forests

Forests have significant impacts on the hydrological cycle. They are responsible for much interception, which in turn means reduced infiltration and overland flow. Forests, of course, are characterised by high levels of transpiration.

Like wetlands, drought threatens forests, but it is people and deforestation that most threaten their survival. In the coniferous forests, drought is not only causing direct physiological damage but is also increasing the susceptibility of pines and firs to fungal diseases. Tree mortality is on the increase. The same applies to the tropical rainforest, except that the increased mortality attributed to drought appears to be having a greater impact on large trees. Here there is the added concern of what this increased tree mortality will eventually do to this incredibly important global carbon store (see p. 139).

As ecosystems play such a vital role within the hydrological cycle, it is important to ensure that their ecological resilience is not overstretched by either the destructive activities of people or natural events such as droughts and floods.

Grey water is waste bath, shower, sink and washing water. It can be recycled, including for human consumption.

Ecological resilience is the capacity of an ecosystem to withstand and recover from a natural event (such as drought and flooding) or from some form of human disturbance.

Exam tip

Wetlands are a prime example of an undervalued ecosystem. You should understand the reasons why.

Surpluses within the hydrological cycle

The physical causes of flooding

Surpluses within the hydrological cycle more often than not mean flooding. The meteorological causes of flooding are:

- intense storms which lead to **flash flooding**, as in semi-arid areas but more commonly in mountainous areas
- prolonged, heavy rain, such as during the Asian monsoon and with the passage of deep depressions across the UK
- rapid snowmelt during a particularly warm spring, as on the plains of Siberia.

Bangladesh is a particularly flood-prone country mainly because it is a land of floodplains and deltas built up by mighty rivers such as the Ganges, Padma and Meghna (Figure 61). These rivers are swollen twice a year by meltwater from the Himalayas and by the summer monsoon. Hilly tracts between the rivers and behind Chittagong are often victims of flash floods.

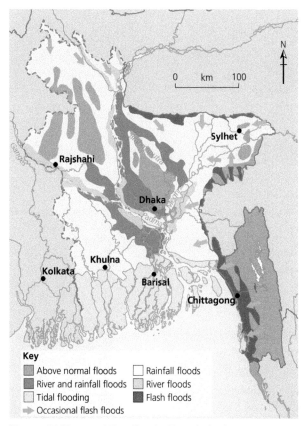

Key
- ⬛ Above normal floods
- ⬜ River and rainfall floods
- ⬜ Tidal flooding
- ➡ Occasional flash floods
- ⬜ Rainfall floods
- ⬜ River floods
- ⬛ Flash floods

Figure 61 Types of flooding in Bangladesh

Figure 61 reveals yet another type of flooding — tidal flooding — often resulting from **storm surges** or when high river flows meet particularly high spring tides in estuaries. The likelihood of flooding is also increased by other physical circumstances:

- in low-lying areas with impervious surfaces, as in towns and cities
- where the ground surface is underlain by impermeable rocks

Flash flooding is distinguished by its exceptionally short lag-time — often minutes or hours.

Knowledge check 69

Suggest reasons why flash floods often occur in mountainous areas.

A **storm surge** is caused by very low air pressure which raises the height of the high-tide sea. Strong onshore winds then drive the 'raised' sea towards the coast, often breaching coastal defences and flooding large areas.

- when ice dams suddenly melt and the waters in glacial lakes are released
- where volcanic activity generates meltwater beneath ice sheets that is suddenly released (jökulhlaups)
- where earthquakes cause the failure of dams or landslides that block rivers.

Knowledge check 70

How else do plate tectonics contribute to flooding?

Human activity and the flood risk

A combination of economic and population growth during the 20th century caused many floodplains to be built upon and many natural landscapes to be modified for agricultural, industrial and urban purposes. The impacts of human activities on the hydrological cycle were examined on p. 94. These same activities, all related to changing land use within river catchments, frequently increase the flood risk (Figure 62), none more so than urbanisation (see p. 100).

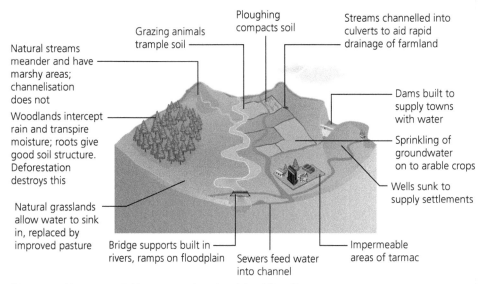

Figure 62 Human activities increasing the risk of flooding

One human activity not explicitly shown in Figure 62 passes under the umbrella title of river mismanagement. For example:

- channelisation: an effective way of improving river discharge and reducing the flood risk. The trouble is that it simply displaces that risk downstream. Some other location may well be overwhelmed by the increased discharge
- dams: block the flow of sediment down a river so the reservoir gradually fills up with silt; downstream there is increased river bed erosion
- river embankments: designed to protect from floods of a given magnitude. They can fail when a flood exceeds their capacity. Inevitably, when this happens, the scale of the flooding is that much greater.

These examples of hard-engineering intervention serve as reminders that soft-engineering methods of reducing the flood risk are preferable. These include making greater use of floodplains as nature intended, namely as temporary stores of flood water, and using them only for nature conservation and perhaps agriculture and recreation.

The impacts of flooding

Socioeconomic

The impacts of flooding are all too familiar. They include:

- death and injury
- spread of water-borne diseases
- trauma
- damage to property, particularly housing
- disruption of transport and communications
- interruption of water and energy supplies
- destruction of crops and loss of livestock
- disturbance of everyday life, including work.

Environmental

The environmental impacts of flooding receive much less publicity. Perhaps it is because there are some positives, which include:

- recharged groundwater stores
- increased connectivity between aquatic habitats
- soil replenishment
- for many species, flood events trigger breeding, migration and dispersal.

Most ecosystems have a degree of ecological resilience that can cope with the effects of moderate flooding. It is where the environment has been degraded by human activities that negative impacts are more evident. For example, the removal of soil and sediment by floodwaters can lead to the eutrophication of water bodies. That same floodwater can also leach pollutants into water courses with disastrous effects for wildlife, while diseases carried by floodwater can weaken or kill trees.

UK floods 2007–2020

The UK has experienced numerous severe floods in recent years, most notably in the summers of 2007 and 2012 and the winters of 2013, 2015–2016 and 2019–2020 (Figure 63 on p. 108).

Knowledge check 71

What is the difference between hard- and soft-engineering approaches to flood control?

Exam tip

Have some key facts about a particular flood event ready. They will add conviction to your answers.

Eutrophication is the process of nutrient enrichment that ultimately leads to the reduction of oxygen in rivers, lakes and ponds, and the consequent death of fish and other species.

Exam tip

It is important to remember that the impacts of flooding are not all negative.

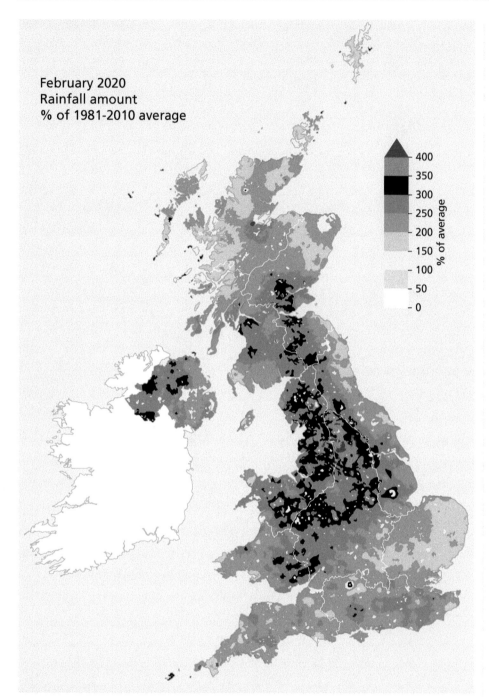

February 2020
Rainfall amount
% of 1981-2010 average

Figure 63 Rainfall in the UK, February 2020
Source: Met Office

These severe floods have had the same basic cause, namely prolonged heavy rainfall, but at different times of the year. During the February 2020 floods, large areas of the UK received 250% of the average amount of rainfall for that time of the year.

Worcestershire and parts of Yorkshire were among the worst-hit places and were the focus of media attention. These floods led to 11 deaths and at least £150 million in property damage.

As is so often the case after flood events, there were recriminations about the apparent inadequacy of flood protection measures. The following were singled out for blame:

- budget cuts in the amount of money being spent on flood defences
- environmental conservation taking priority over the regular dredging of rivers
- poor land management, resulting in blocked ditches
- global warming.

What tends to be forgotten in post-flood enquiries is that flood protection measures are designed to cope with flood events of a given magnitude. When an event of a very rare order of magnitude occurs, no amount of money or engineering is going to provide the hoped-for degree of protection.

Exam tip

Be cautious about the possible link between increasing frequency of high-magnitude floods in the UK and global warming, as the link between them is not yet proven.

Climate change and the hydrological cycle

Impacts of climate change on inputs and outputs

Much research has yet to be done to establish beyond any doubt the impact of global warming and climate change on the hydrological cycle. Table 54 summarises the findings to date concerning the inputs and outputs.

Knowledge check 72

What is significant about soil moisture?

Table 54 Summary of impacts of climate change on inputs and outputs

Precipitation	<ul><li>A warmer atmosphere has a greater water-holding capacity</li><li>Widespread increases in rainfall intensity are expected more than large increases in total amounts</li><li>Areas of precipitation increase include the tropics and high latitudes</li><li>Areas of precipitation decrease lie between 10° and 30° north and south of the Equator</li><li>The length and frequency of heatwaves are increasing in some locations and are resulting in the increased occurrence of drought</li><li>With climate warming, more precipitation in northern regions is falling as rain rather than snow</li></ul>
Evaporation and evapotranspiration	<ul><li>Evaporation over large areas of Asia and North America appears to be increasing</li><li>Transpiration is linked to vegetation changes, which in turn are linked to changes in soil moisture and precipitation</li></ul>
Soil moisture	<ul><li>Impact is uncertain, as soil moisture depends on many factors, of which climate is only one</li><li>Where precipitation is increasing, it is likely that soil moisture will also increase</li></ul>

Impacts of climate change on stores and flows

Table 55 (p. 110) summarises some possible changes. There is less certainty here compared with Table 54.

Concerns about short-term oscillations (ENSO cycles)

One of the problems with this forecasting of possible changes to the hydrological cycle is distinguishing between the impacts of long-term climate change and those of the short-term oscillations associated with El Niño events. A further complication results from the fact that ENSO cycles are associated with both extreme flooding in some parts of the world and extreme drought in others.

Exam tip

Remember that some of the impacts in Tables 54 and 55 are possible rather than probable.

Table 55 Summary of impacts of climate change on flows and stores

Surface runoff and stream flow	■ More low flows (droughts) and high flows (floods) ■ Increased runoff and reduced infiltration
Groundwater flow	■ Uncertain because of abstraction by humans
Reservoir, lake and wetland storage	■ Changes in wetland storage cannot be conclusively linked to climate change ■ It appears that storage is decreasing as temperatures increase
Soil moisture	■ Possibly little change, with higher precipitation and evaporation cancelling each other out
Permafrost	■ Deepening of the active layer is releasing more groundwater ■ Methane released from thawed lakes may be accelerating change
Snow	■ Decreasing length of snow-cover season ■ Spring melt starting earlier ■ A decreasing temporary store
Glacier ice	■ Strong evidence of glacier retreat and ice sheet thinning since the 1970s ■ Less accumulation because more precipitation falling as rain ■ A decreasing store
Oceans	■ More data on surface temperatures needed ■ Where there is ocean warming, there will be more evaporation ■ Possibly ocean warming leads to the generation of more cyclones ■ Storage capacity being increased by meltwater ■ Rising sea level

Knowledge check 73

Which store, in Table 55, is expected to increase in volume?

What is perhaps of more concern is the potential impact of short-term climate change (regardless of whether or not it is related to ENSO cycles) on global water supplies. Figure 64 suggests a scenario of increased uncertainty about the security of water supplies (for more, see the next section).

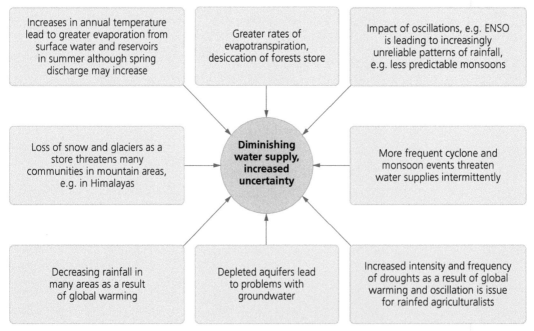

Figure 64 The impacts of short-term climate change on water supply

Figure 64 reminds us that we live in an uncertain world. Here the uncertainty arises simply because even modern scientific research is unable to make confident forecasts about the future availability of water. But even if scientists were able to do this, there are other important unknowns to be taken into account, such as possible advances in water technology and factors related to the demand side, such as population growth and the rising tide of global development.

How does water insecurity occur and why is it becoming such a global issue for the 21st century?

- Water insecurity is a major concern for many countries and is the outcome of both physical and human factors.
- Water insecurity brings with it both serious consequences and considerable risks.
- There is a growing need for more sustainable management of water supplies. This can be achieved through a number of approaches.

The causes of water insecurity

The growing mismatch between water supply and demand

It is important to start by recalling the message conveyed by Figure 52 (p. 91), namely that accessible surface freshwater is a scarce resource. There is increasing pressure on that resource largely as a result of population growth and economic development. The situation is that many countries are experiencing **water insecurity**. Figure 65 shows the global distribution of renewable water resources.

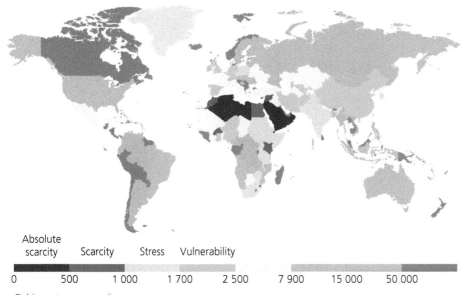

Absolute scarcity | Scarcity | Stress | Vulnerability

0   500   1 000   1 700   2 500   7 900   15 000   50 000

Cubic metres per capita per year

Figure 65 The global distribution of renewable water resources

The growing mismatch relates to the distribution of freshwater resources (water availability) and the distribution of the demand for water. Unfortunately, they do not coincide; far from it. This is where water insecurity begins. So let us focus first on these contributory factors: water availability on the one hand and the rising demand for water on the other.

Water insecurity begins to exist when available water is less than 1700 m³ per person per day. This marks the start of what is known as water stress. Below 1000 m³ per person per day, water stress gives way to water scarcity (Figure 65).

Water availability

Water availability is caused initially by climate, i.e. by annual precipitation. That water is then moved and distributed by the drainage network. But there are a number of factors that reduce the amount of water that is eventually available for human use. These include both human and physical factors:

- evaporation and evapotranspiration
- discharge into the sea
- saltwater encroachment at the coast
- contamination of water by agricultural, industrial and domestic pollution
- over-abstraction from rivers, lakes and aquifers and the acute need to replenish these dwindling stores.

Because of these human factors, it is appropriate to talk of a diminishing water supply. The situation is also being exacerbated by global warming and climate change (see Figure 64 on p. 110.

Rising water demand

The rising demand for water is driven by three main factors:

- population growth: more people, more thirsts to quench
- economic development: increases the demand for water in almost all economic activities — agriculture, industry, energy and services. One of the biggest and fastest-growing consumers is irrigation (see more on p. 114)
- rising living standards: increase in the per capita consumption of water for drinking, cooking, bathing and cleaning. Added to this domestic consumption are water-extravagant items such as swimming pools, washing machines and dishwashers.

An important point here is that within the rising demand for water, there is increasing competition between water users for this dwindling resource. It is becoming increasingly serious in some locations. Figure 66 shows the three main pressures that are increasing the risk of water insecurity.

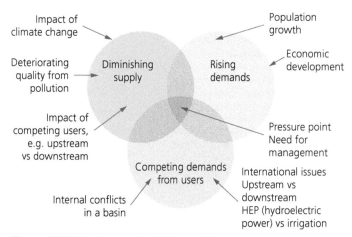

Figure 66 Water insecurity in the making

> **Knowledge check 75**
>
> Suggest a definition for the term 'water availability'.

> **Exam tip**
>
> Be sure you understand how these three drivers increase the demand for water.

> **Knowledge check 76**
>
> Study Figure 66. What is meant by 'upstream vs downstream'?

> **Synoptic link**
>
> (F): How many people will suffer water scarcity in the future? This is an important question but a difficult one to answer. As water is so fundamental to human life, widespread water scarcity could be a major risk.

The consequences and risks of water insecurity
Water and economic scarcity

A distinction was made above between water stress and water scarcity. There is a further distinction to be made in water scarcity: that between physical scarcity and economic scarcity.

Physical scarcity occurs when more than 75% of a country's or region's **blue water** flows are being used. This currently applies to about 25% of the world's population. Qualifying countries are located in the Middle East and North Africa (Figure 67). Qualifying regions occur in north China, western USA and southeast Australia.

Economic scarcity occurs where the use of blue water sources is limited by lack of capital, technology and good governance so some people cannot afford an adequate water supply. It is estimated that around 1 billion people are restricted from accessing blue water by high levels of poverty. Most of these people live in Africa and South Asia (Figure 67).

Blue water is water stored in rivers, streams, lakes and groundwater in liquid form.

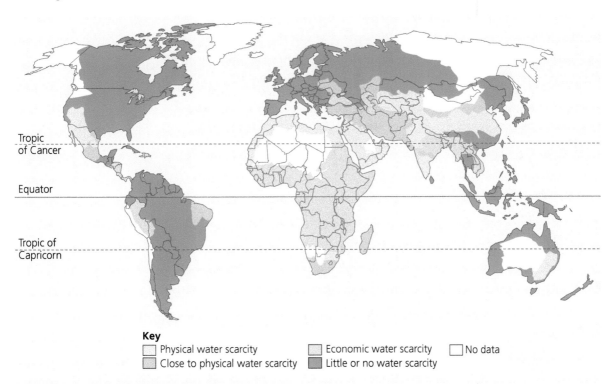

Key

☐ Physical water scarcity
☐ Close to physical water scarcity
☐ Economic water scarcity
■ Little or no water scarcity
☐ No data

Figure 67 The global distribution of water scarcity

In short, the causes of water scarcity are twofold:

1 a lack of precipitation, either annually or seasonally

2 a lack of the ability to harness the amount of blue water in demand.

Exam tip

The examiner may be impressed if you show that you know about these two types of water scarcity.

Access to **safe water** is regarded by some as a human right. In the 21st century, however, it is increasingly seen as a commodity for which a realistic price should be paid:

■ This might be all well and good in the developed world. Indeed, much of the water supply industry there is now in the hands of private companies. People expect to have to pay for their water.

■ In the developing world, however, the situation is very different. In the UK, the cost of 50 litres of tap water is about 10p, but in Accra in Ghana it is about 50p from a water tanker. That 50p is 20–25% of an informal worker's daily income. Supplying safe water in areas of physical water scarcity can be difficult, costly and well beyond the means of very poor people. This is where charities such as WaterAid provide such invaluable help. Their programmes are helping to reduce the extent of economic water scarcity.

Safe water is water fit for human consumption.

Exam tip

The price of water varies from place to place depending on the availability and the level of demand. It is when demand exceeds supply that the price rises.

Water supply and economic development

The point has already been made that economic development is one of the main drivers of the increasing demand for water.

Agriculture

Figure 68 shows the extent to which agriculture dominates water use. This is not surprising because around 20% of the world's land is under full irrigation. Around 30% of this irrigation comes from dams and their networks of irrigation canals. But the majority of irrigation water is pumped from aquifers and is leading to massive groundwater depletion, especially in China, India, Pakistan and the USA. Clearly, this water situation is unsustainable and hydrological cycles are being seriously disrupted.

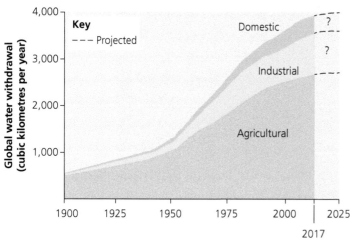

Figure 68 Trends in water use

Industry and energy

Just over 20% of all freshwater withdrawals worldwide are for industrial and energy production. Industries such as chemicals, electronics, paper, petroleum and steel are major consumers of water. Water pollution is a significant problem associated with much of this industrial use of water.

Over half of the water used by energy production is either for generating **HEP** or as cooling water in thermal and nuclear power stations. So all this water is returned to its source virtually unchanged. However, there is mounting concern about the growth of biofuels for the production of bioethanol and biodiesel. Unfortunately these crops are very thirsty.

HEP is the abbreviation for hydroelectric power.

Domestic use

With economic development come rising standards of living and an increasing per capita consumption of water. Safe water is a fundamental human need. However, water does have its risks as far as human wellbeing is concerned.

Water, particularly that polluted by a lack of sanitation, is an effective medium for the breeding and transmission of a range of lethal diseases, such as typhoid, cholera and dysentery. Water is also a productive breeding ground for some disease vectors, such as mosquitoes, snails and parasitic worms. Malaria, dengue fever and bilharzia are debilitating vector diseases. So safe water is vital to human health, particularly in the context of washing and food preparation.

Knowledge check 77

Why do rising standards of living lead to an increase in per capita water consumption?

From the above, it will be understood that an inadequate supply of water can easily impede any water-dependent aspects of economic development. Costs may well rise. An inadequate water supply will also threaten human health. Environmentally, it will encourage people to over-exploit what water resources there are. This could easily prolong periods of drought and possibly be a first step on the downward path to desertification.

Potential water conflicts

When the demand for water overtakes the available supply and there are key stakeholders desperate for that water, there is potential for conflict, or what have been called 'water wars':

- Within countries, conflicts can arise between the competing demands of irrigation, energy, industry, domestic use and recreation.
- But it is when countries 'share' the same river or drainage basin, as is the case with transboundary water sources, that the 'normal' competition for water can be raised to a different level, namely one of international tensions and even open conflict. Serious conflict over water supply usually occurs only when countries have other disagreements as well, such as border disputes.

Exam tip

Take care when using the world 'conflict'. In most instances, it means 'disagreement' rather than armed conflict. 'Tension' is a good alternative.

The Nile is a truly remarkable river. At 6700 km, it is the world's longest river. Even more remarkable is the fact that no less than 11 countries compete for its water. Currently 300 million people live within the Nile basin and such is the rate of population growth that that total is set to double by 2030.

All these people will need the waters of the Nile for domestic consumption and for growing crops. The Nile is also expected to generate HEP. Potential flash points have been the dams and barrages built in Sudan and Ethiopia that deprive downstream Egypt of its fair share of Nile water. Both Sudan and Egypt have serious concerns over the opening of Ethiopia's Grand Ethiopian Renaissance Dam (GERD), completed in 2020, and its impact on downstream water supply. Other shared rivers that could become the battlefields of water wars are the Jordan and the Tigris–Euphrates in the Middle East, and the Indus and Ganges in the Indian subcontinent.

Knowledge check 78

Name the countries involved in each of these four shared rivers: Jordan, Tigris–Euphrates, Indus, Ganges.

Synoptic link

(P): Players often come into conflict over water supply, ranging from minor disputes to near wars at scales from local to international. At a local scale, key players are the water users (farmers, industrialists and households). Their views may well differ from those of planners, environmentalists and water providers. Internationally, the key players are those governments and users of transboundary water sources. In some case, it may be necessary to call in the mediating services of UN agencies.

Different approaches to managing water supply

Because of the wide range of players involved in the use of water resources, there are inevitably conflicting views over what constitutes the best approach to the management of those resources. For example, economic players, such as businesses, typically opt for hard-engineering schemes, while environmental players, such as conservation organisations, favour a more sustainable approach.

Hard-engineering schemes

These require high levels of capital and technology. There are now up-and-running long-distance water transfer schemes, mega dams and clusters of desalinisation plants.

Water transfers

Water transfer schemes involve the diversion of water from one drainage basin to another, either by diverting a river or by constructing a large canal to carry water from one basin to another.

Perhaps the most publicised of these is China's South–North Transfer Project which is currently under construction. The idea of moving water from an area of surplus to one of deficit is a deceivingly simple one, but as Figure 69 shows, there are issues.

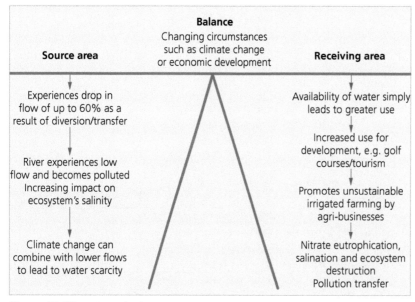

Figure 69 Water transfer issues

Mega dams

Nearly 60% of the world's major rivers are impeded by large dams, perhaps most notably the Colorado, Nile and Yangtze. While the capital costs of such dams are immense, this hard-engineering solution to water shortages has other drawbacks, such as the high evaporation losses from the water surface, the disruption of the downstream transport of silt and the displacement of people.

Desalinisation

Given the increasing pressure on freshwater resources and their shortage in the drier parts of the world, it is not surprising that some countries are looking to the oceans for water. Desalinisation has been undertaken on a small scale for centuries, but recently there have been technological advances in the process, most notably:

- development of the process of reverse osmosis
- pioneering work on carbon nanotube membranes.

Desalinisation is an expensive process; it requires inputs of advanced technology and energy. However, as the price of freshwater rises, so some countries will look increasingly to the seas for their water supplies. Some Middle Eastern states, such as Saudi Arabia, Kuwait and UAE, have already done so.

Because of its inputs, desalinisation is classified as a hard-engineering solution. However, it is a sustainable process, although it does have an ecological impact on marine life.

Desalinisation is the process by which dissolved solids in sea water are partially or completely removed to make it suitable for human use.

Knowledge check 79

Why is desalinisation a limited option for increasing global water supplies?

Sustainable water management

The main aims of sustainable water managements are to:

- minimise wastage and pollution of water resources
- ensure that there is access to safe water for all people at an affordable price
- take into account the views of all water users
- guarantee an equitable distribution of water within and between countries.

There is no silver bullet here. Rather, a diversity of actions is being taken today as steps towards those four management aims. They include the following:

- Smart irrigation: a top priority given the huge amount of water used by irrigation. Traditional sprinkler and surface flow systems are being replaced by modern automated spray technology and advanced drip irrigation systems.
- Hydroponics: growing crops in greenhouses that are carbon dioxide and temperature controlled in shallow trays where they are drip-fed nutrients and water; there is no soil.
- Recycling of grey water: a low-cost option that produces water for agricultural use but not human consumption.
- Rainwater harvesting: people collect the rain falling on the roofs of dwellings and store it in butts for various domestic purposes, such as flushing toilets and watering the garden.
- Filtration technology: this is now so effective that there is little dirty water that cannot be physically purified and recycled.
- Restoration: of damaged rivers, lakes and wetlands so that they can play their full and proper part in the hydrological cycle.

These and other actions are clearly environmentally sustainable and can bring many socioeconomic benefits to local communities. The only question is: are they economically sustainable?

Singapore

Circumstances — few natural water resources, a thriving economy, a high standard of living and a high per capita consumption of water — have made water management a top priority in this tiny state with its nearly 6 million inhabitants. It has adopted a holistic approach to water management based on three key strategies:

1 Collect every drop of water: the government has various ways of encouraging citizens to use water prudently. Since 2003, per capita domestic water consumption has fallen from 165 litres per day to 141 litres per day.

2 Re-use water endlessly: Singapore is at the cutting edge of new technologies to re-use grey water (called NEWater).

3 Desalinate more seawater: five desalinisation plants now meet up to 25% of the water demand.

Despite these impressive actions, Singapore has to import water from neighbouring Malaysia.

Integrated drainage basin management (IWRM)

IWRM (Integrated Water Resources Management) was first advocated in the late 1990s. It emphasises the river basin as a logical geographical unit for the management of water resources. It is based on achieving a close co-operation between basin users and players. The river basin is treated holistically in order to ensure three things:

■ the environmental quality of the rivers and catchment
■ that water is used with maximum efficiency
■ an equitable distribution of water among users.

Experience has shown that IWRM works well at a community level but not so well in larger river basins, especially if an international boundary is involved, as with the Colorado River and the Nile.

Water-sharing treaties and frameworks

In spite of the potential for conflicts over shared waters, particularly where there is 'greedy' upstream behaviour, international co-operation is the rule rather than the exception. Over the last 60 years, military conflict has occurred in only a handful of drainage basin disagreements. There has been a surprising amount of international co-operation, even between traditional rivals, as for example between India and Pakistan which share the Indus.

Important international agreements include:

■ the Helsinki Rules (1966) and Berlin Rules (2004) with their 'equitable use' and 'equitable shares' concepts
■ the United Nations Economic Commission for Europe (UNECE) Water Convention which promotes the joint management and conservation of shared freshwater ecosystems
■ the UN Water Courses Convention which offers guidelines for the protection and use of transboundary rivers

Exam tip

You will need to show that you are aware of some of these sustainable actions.

Synoptic link

(A): Attitudes to water use and supply vary. Some players view safe water as a human right, whereas politicians see it as a human need which they have to supply. Businesses will secure their water needs almost regardless of the costs, whereas environmentalists argue provision should be sustainable.

Exam tip

As yet there are few well-publicised examples of IWRM. It is a neat idea, but …

- the EU Water Framework Directive (2000), committing all members to ensure the 'status' of their water bodies, including their marine waters up to one nautical mile from shore.

In short, the potential for water wars is considerable, and increasing with climate change. However, a commendable degree of international co-operation seems to be keeping the peace. But for how long?

Synoptic link

(P): Numerous players are involved in water conflicts, all with different demands. This makes agreement over management and supply very challenging, and is one reason why there are regional and international mechanisms to help broker agreement. The breakdown of agreements brings very high risks, as no one can live without water.

Summary

- The global hydrological cycle is of enormous importance to all life on Earth. It is a closed system of stores and flows and is driven by solar energy and gravity. The global water budget determines the amount of water that is available for human use.
- The drainage basin is an open system within the global hydrological cycle. The system is often disrupted by human activities, such as deforestation, agriculture and urbanisation.
- Understanding water budgets, river regimes and storm hydrographs is important in the proper management of the land and water resources of drainage basins.
- Hydrological systems are constantly varying over a range of timescales. Recurrent droughts and floods are reminders of short-term variations or anomalies.
- In both instances, human activities have the potential to turn the deficits and surpluses of precipitation into hazards and sometimes disasters.
- The occurrence of droughts and floods is usually explained by scientists in terms of short-term weather patterns and medium-term climate cycles (ENSO). However, the increasing frequency and scale of these events is leading scientists to now look for explanations in much longer-term climate change, i.e. global warming.

- Climate change is undoubtedly already affecting the inputs and outputs of hydrological systems, as well as their stores and flows.
- Water insecurity arises out of a mismatch between the distributions of water supply and water demand. Its causal factors are both physical and human. An increasing demand for water is being driven by population growth, economic development and rising living standards. At the same time, the availability of water is being diminished by global warming and human abuse.
- Given that freshwater is a scarce but vital commodity, there is considerable potential for conflicts to occur between users within a country, and internationally over transboundary water sources.
- Proper management of water supply is crucial. It can be undertaken in a number of ways. Some, such as hard-engineering schemes, are much less sustainable than water recycling or integrated drainage basin management schemes.
- Given that there are many 'shared' rivers, it is important that there are in place binding international agreements.

■The carbon cycle and energy security

How does the carbon cycle operate to maintain planetary health?

■ Most of the world's carbon is locked away in terrestrial stores as part of a long-term geological cycle.
■ On shorter timescales, biological processes sequester carbon both on land and in the oceans.
■ A balanced carbon cycle is important in sustaining other Earth systems, but the balance is being increasingly upset by human activities.

Terrestrial carbon stores

Carbon is everywhere: in the oceans, in rocks and soils, in all forms of life and in the atmosphere. Without carbon, life would not exist as we know it. The wellbeing and functioning of the Earth depends on carbon and how it cycles through the Earth's systems.

Stores and fluxes

Figure 70 shows the **carbon cycle** and its two main components: **stores** and **fluxes**.

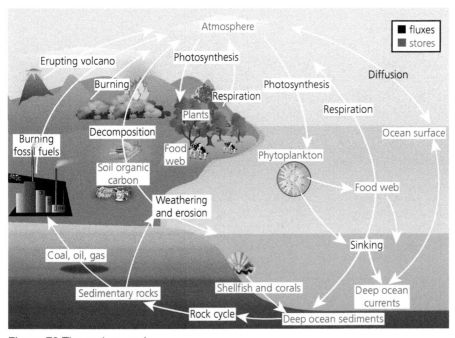

Figure 70 The carbon cycle

The **carbon cycle** is the cycle by which carbon moves from one Earth sphere (atmosphere, hydrosphere, lithosphere and biosphere) to another. It is a closed system but is made up of interlinked subsystems which are open and have inputs and outputs.

Carbon stores function as sources (adding carbon to the atmosphere) and sinks (removing carbon from the atmosphere).

Carbon fluxes (also known as flows) are movements of carbon from one store to another; they provide the motion in the carbon cycle.

Carbon exists in different forms, depending on the store:

- atmosphere: as carbon dioxide (CO_2) and carbon compounds, such as methane (CH_4)
- hydrosphere: as dissolved CO_2
- lithosphere: as carbonates in limestone, chalk and fossil fuels, as pure carbon in graphite and diamonds
- biosphere: as carbon atoms in living and dead organisms.

These stores vary in size and capacity, as well as in location. The important distinction in the biosphere is between terrestrial and oceanic locations.

Carbon fluxes, or flows, between the carbon stores of the carbon cycle are measured in either petagrams or gigatonnes of carbon per year. The major fluxes are between the oceans and the atmosphere, and between the land and the atmosphere via the biological processes of photosynthesis and respiration. These fluxes vary not only in terms of their rates of flow but also on different timescales.

> **Exam tip**
>
> Use key words relating to the carbon cycle (such as store, flux, sink, pump, sequestration) in your answers.

Geological origins

Most of the Earth's carbon is geological and results from:

- the formation of sedimentary carbonate rocks (limestone) in the oceans. The Himalayas form one of the Earth's largest carbon stores. This is because the mountains started life as ocean sediments rich in calcium carbonate derived from shells, corals and plankton. Since these sediments have been uplifted by tectonic processes, the carbon they contain has been weathered, eroded and transported back to the oceans
- carbon derived from plants and animals in shale, coal and other rocks. These rocks were made up to 300 million years ago from the remains of organisms. These remains sank to the bottom of rivers, lakes and seas and were subsequently covered by silt and mud. The remains continued to decay anaerobically and were compressed and heated, eventually forming coal, oil and natural gas. The subsequent burning of these fossil fuels has, of course, released the large amounts of carbon they contained back into the atmosphere.

Geological processes releasing carbon

The release of geological carbon into the atmosphere results not just from people burning fossil fuels but also through two natural processes:

1 Carbon dioxide in the atmosphere reacts with moisture to form weak carbonic acid. When this falls as rain, it reacts with some of the surface minerals and slowly dissolves them, i.e. there is **chemical weathering**.

2 Pockets of carbon dioxide exist in the Earth's crust. Volcanic eruptions and earthquakes can release these gas pockets. This **outgassing** occurs mainly along mid-oceanic ridges, subduction zones and at magma hotspots.

> **Exam tip**
>
> Be sure you know where carbon is stored and in what form.

> **Knowledge check 80**
>
> Name one carbon store and one carbon flux (or flow).

Chemical weathering is the decomposition of rock minerals in their original position by agents such as water, oxygen, carbon dioxide and organic acids.

Outgassing is the release of gas previously dissolved, trapped, frozen or absorbed in some material (e.g. rock).

Biological processes sequestering carbon

Compared with its geological counterpart, biological **sequestrating** operates on much shorter timescales — from hours to centuries.

Oceanic sequestering

The oceans are the Earth's largest carbon store. The oceanic store of carbon is 50 times greater than that of the atmosphere. Most of the oceanic carbon is stored in marine algae, plants and coral. The rest occurs in dissolved form.

There are three types of oceanic **carbon pump** (Figure 71).

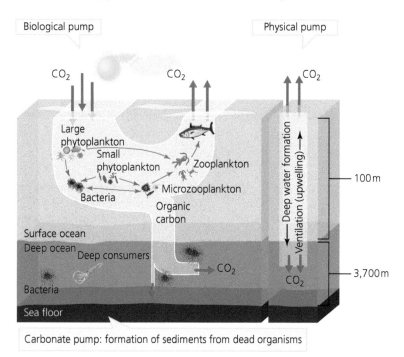

Figure 71 Oceanic carbon pumps

1   The biological pump moves carbon dioxide from the ocean surface to marine plants (**phytoplankton**) by a process known as **photosynthesis**. This effectively converts carbon dioxide into food for zooplankton (microscopic animals) and their predators. Most of the carbon dioxide taken up by phytoplankton is recycled near the surface. About 30% sinks into deeper waters before being converted back into carbon dioxide by marine bacteria.

2   The physical pump moves carbon to different parts of the ocean in downwelling and upwelling currents. Downwelling occurs in those parts of the oceans where cold, denser water sinks. These currents bring dissolved carbon dioxide down to the deep ocean. Once there, it moves in slow-moving deep ocean currents, staying there for hundreds of

Carbon sequestration is the process by which carbon dioxide is removed from the atmosphere and held in solid or liquid form. It facilitates the capture and storage of carbon.

Carbon pumps are the processes operating in the oceans that circulate and store carbon.

Phytoplankton consists of the microscopic plants and plant-like organisms drifting or floating in the sea (or freshwater) along with diatoms, protozoa and small crustaceans.

Photosynthesis is the process by which green plants and some other organisms use sunlight to synthesise (extract) nutrients from carbon dioxide and water.

years. Eventually, these deep ocean currents, part of the **thermohaline circulation**, return to the surface by upwelling. The cold deep ocean water warms as it rises towards the ocean surface and some of the dissolved carbon dioxide is released back into the atmosphere.

3 The carbonate pump forms sediment, and eventually rock, from dead organisms that fall to the ocean floor, including the hard outer shells and skeletons of shells, fish and corals.

Figure 72 shows the thermohaline circulation, a giant conveyor belt that moves water of varying temperatures, densities and salinities through the oceans. As a consequence, it plays a vital part in the carbon cycle.

The **thermohaline circulation** is the global system of surface and deep ocean currents driven by temperature (thermo) and salinity (haline) differences between various parts of the oceans (Figure 72).

Knowledge check 81

Name the three types of oceanic carbon pump.

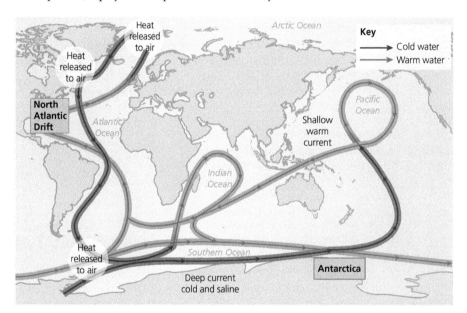

Figure 72 The thermohaline circulation

Terrestrial sequestering

Plants (i.e. primary producers in the ecosystem) sequester carbon out of the atmosphere during photosynthesis. In this way, carbon enters the food webs and nutrient cycles of terrestrial ecosystems (see Figure 70 on p. 120):

■ When animals eat plants, carbon sequestered in the plant becomes part of their fat and protein.

■ Respiration by animals returns some of the carbon back to the atmosphere as exhaled CO_2.

■ Waste from animals is eaten by micro-organisms (bacteria and fungi) and detritus feeders (e.g. beetles).

■ As a consequence, carbon becomes part of these creatures.

■ When plants and animals die and their remains fall to the ground, carbon is released into the soil.

Carbon fluxes within ecosystems vary on two timescales:

1 diurnally: during the day, fluxes are positive — that is, from the atmosphere into the ecosystem; at night the reverse situation applies

2 seasonally: during winter, carbon dioxide concentrations increase because of the low levels of plant growth. However, as soon as spring arrives and all plants grow again, those concentrations begin to decrease until the onset of autumn.

Biological carbon

All living organisms contain carbon; the human body is about 18% carbon by weight. In plants, carbon dioxide and water are combined to form simple sugars, i.e. carbohydrates. In animals, carbon is synthesised into complex compounds, such as fats, proteins and nucleic acids.

On land, soils are the biggest carbon stores. Here biological carbon is stored in the form of dead organic matter. This matter can be stored for decades or even centuries before being broken down by soil microbes (biological decomposition) and then either taken up by plants or released back into the atmosphere.

Soils store between 20% and 30% of global carbon. They sequester about twice the quantity of carbon as the atmosphere and three times that of terrestrial vegetation. The actual amount of carbon stored in a soil depends on:

- climate: this dictates the rates of plant growth and decomposition — both increase with temperature and rainfall
- vegetation cover: this affects the supply of dead organic matter, being heaviest in tropical rainforests and least in the tundra
- soil type: clay protects carbon from decomposition, so clay-rich soils have a higher carbon content
- land use: cultivation and other forms of soil disturbance increase the rate of carbon loss.

Increasing human interference

A fully functioning and balanced carbon cycle is vital to the health of the Earth in sustaining its other systems. It plays a key role in regulating the Earth's temperature by controlling the amount of carbon dioxide in the atmosphere. This, in turn, affects the hydrological cycle. Ecosystems, terrestrial and oceanic, also depend on the carbon cycle. All this is a consequence of the fact that the carbon cycle provides the all-important link between the atmosphere, hydrosphere, lithosphere and biosphere. But the carbon balance is being increasingly altered by human actions and activities.

The greenhouse effect

It is the increasing concentration of carbon in the atmosphere that is causing concern. Carbon dioxide and methane are perhaps the most important of all the **greenhouse gases (GHGs)**. Their increasing presence in the atmosphere is upsetting the Earth's natural temperature-control system, resulting in an enhanced greenhouse effect (Figure 73) where the natural greenhouse effect works more powerfully than it should, warming the planet.

Exam tip

Make sure you know the differences between sequestering and photosynthesis.

Exam tip

Remember that soils both sequester and release carbon dioxide. The balance between the two fluxes varies with local conditions, such as climate, soil type, vegetation cover and land use.

Knowledge check 82

What factors affect the amount of carbon stored in the soil?

Greenhouse gases (GHGs) are primarily water vapour, carbon dioxide, methane, nitrous oxide and ozone. These gases both absorb and emit solar radiation and in so doing create the so-called greenhouse effect that determines global temperatures.

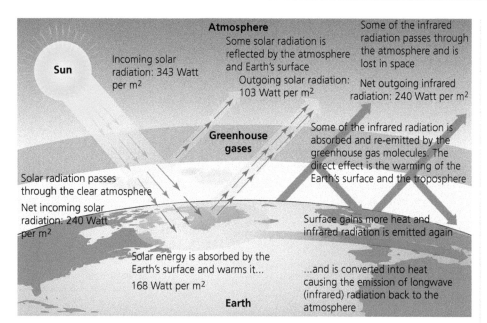

Figure 73 The greenhouse effect

The Earth's climate is driven by incoming short-wave solar radiation:

- 31% is reflected back into space by clouds, GHGs and the land surface.
- The remaining 69% is absorbed at the Earth's surface, especially by the oceans.
- Much of this radiation absorbed at the surface is re-radiated as long-wave radiation.
- Large amounts of this long-wave radiation are, however, prevented from returning into space by clouds and GHGs.
- The trapped long-wave radiation is then re-radiated back to the Earth's surface.

It is this trapped, re-radiated, long-wave energy that constitutes the natural greenhouse effect and controls the mean global temperature. It also determines the distributions of both heat and precipitation.

Maintaining a balanced carbon cycle

A balanced carbon cycle is the outcome of different components working in a sort of harmony with each other. Here we focus on just two of those components — photosynthesis and soil health.

Photosynthesis

Photosynthesis (see definition on p. 122) by terrestrial and oceanic organisms plays an essential role in keeping carbon dioxide levels relatively constant and thereby helping to regulate the Earth's mean temperature.

The amount of photosynthesis varies spatially, particularly with **net primary productivity** (**NPP**). NPP is highest in the warm and wet parts of the world, particularly in the tropical rainforests and in shallow ocean waters. It is least in the tundra and boreal forests.

Knowledge check 83

What are the main GHGs?

Exam tip

Sometimes it can be quicker and easier in the exam to produce a simplified and annotated diagram of the greenhouse effect.

Knowledge check 84

What helps keep carbon dioxide levels in the atmosphere fairly constant?

Net primary productivity (**NPP**) is the amount of organic matter that is available for humans and other animals to harvest or consume.

Soil health

Soil health is an important aspect of ecosystems and a key element in the normal functioning of the carbon cycle. Soil health depends on the amount of organic carbon stored in the soil. The storage amount is determined by the balance between the soil's inputs (plant and animal remains, nutrients) and its outputs (decomposition, erosion and uptake by plant and animal growth).

Carbon is the main component of soil organic matter and helps to give soil its moisture-retention capacity, its structure and its fertility. Organic carbon is concentrated in the surface layer of the soil. A healthy soil has a large surface reservoir of available nutrients which, in their turn, condition the productivity of ecosystems. All this explains why even a small amount of surface soil erosion can have such a devastating impact on soil health and fertility.

Fossil fuel combustion

Fossil fuels have been burnt to provide energy and power at increasing rates since the beginning of the Industrial Revolution in the mid-18th century. Fossil fuel combustion is the number one threat to the global carbon cycle. It is changing the balance of both the carbon stores and the fluxes.

Knowledge check 85

What are the fossil fuels?

It is estimated that about half the extra emissions of carbon dioxide since 1750 have remained in the atmosphere. The rest have been fluxed from the atmosphere into the stores provided by the oceans, ecosystems and soils. The rate of carbon fluxing has speeded up (see Table 56). It is that additional carbon dioxide in the atmosphere and its impact on the greenhouse effect that is largely responsible for a number of climate changes:

- a rise in the mean global temperature
- more precipitation and evaporation
- sudden shifts in weather patterns
- more extreme weather events, such as floods, storm surges and droughts
- the nature of climate change varying from region to region — some areas are becoming warmer and drier, others wetter.

Table 56 Rising concentrations of atmospheric carbon

	Carbon dioxide (CO_2) concentration (parts per million)	Methane (CH_4) concentration (parts per billion)
Pre-industrial (natural)	280	725
1960	317	1260
1990	354	1725
2020	410	1900

Exam tip

Draw a simple annotated spider diagram summarising some of the impacts of climate change. This type of diagram could be useful in structuring and supporting an exam answer.

These changes in climate have serious knock-on effects on:

- sea level: this is rising because of melting ice sheets and glaciers and thermal expansion as the oceans warm; many major coastal cities around the world are under threat from flooding by the sea
- ecosystems: a decline in the goods and services they provide; a decline in biodiversity; changes in the distributions of species; marine organisms threatened by lower oxygen levels and ocean acidification; the bleaching of corals
- the hydrological cycle: increased temperatures and evaporation rates cause more moisture to circulate around the cycle.

Knowledge check 86

Why is climate change causing the distributions of plant and animal species to change?

What are the consequences for people and the environment of our increasing demand for energy?

- In an energy-hungry modern world, achieving energy security is becoming a top priority for many countries.
- A combination of increasing population, economic development and rising living standards is creating a huge and almost insatiable demand for energy. That demand is being largely met by the burning of fossil fuels.
- There are alternative sources of energy but they all have their costs as well as benefits.

Energy security

Energy security is something that all countries seek to achieve. The most secure energy situation is one where the national demand for energy can be completely satisfied from domestic sources. The more a country depends on imported energy, the more it is exposed to risks of an economic and geopolitical kind. Four key aspects of energy security are (Figure 74):

- availability
- accessibility
- affordability
- reliability.

Energy security is achieved when there is an uninterrupted availability of energy at a national level and at an affordable price.

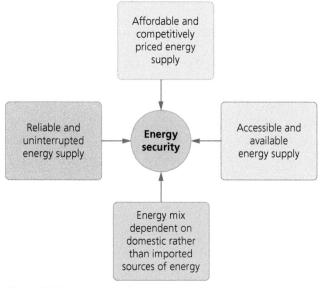

Figure 74 Energy security

The importance of energy security stems from the fact that energy is vital to the functioning of a country. For example, it:

- powers most forms of transport
- lights settlements
- is used by some types of commercial agriculture
- warms or cools homes and powers domestic appliances
- is vital to modern communications
- drives most forms of manufacturing.

Knowledge check 87

What is meant by the 'accessibility of energy'?

Knowledge check 88

What types of commercial farming need a large input of energy?

The energy mix

Figure 75 refers to **energy mix**. It is an important component of energy security. Clearly, the mix or proportions of different energy sources will vary from country to country. Some important distinctions can be made here, for example between:

- domestic and foreign (imported) sources
- **primary** and **secondary energy** sources.

Most energy today is consumed in the form of electricity. The main primary energy sources used to generate electricity are:

- non-renewable fossil fuels, such as coal, oil and natural gas
- recyclable fuels, such as nuclear energy, general waste and biomass
- renewable energies, such as water, wind, solar, geothermal and tidal.

Figure 75 shows how the global primary energy mix has changed since 1900 as energy consumption has dramatically risen.

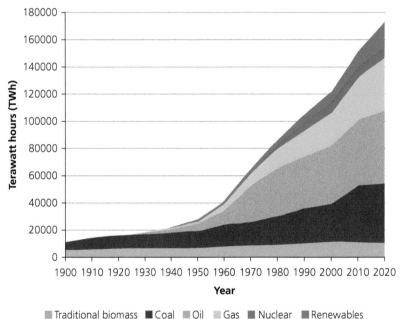

■ Traditional biomass ■ Coal ■ Oil ■ Gas ■ Nuclear ■ Renewables

Figure 75 The changing global energy mix (1900–2020)

Source: BP Statistical Review of World Energy, 2020

The consumption of energy

The consumption of energy is measured in two ways:

1 in per capita terms, i.e. as kilogrammes of oil equivalent or megawatt hours per person. In general, this measure rises with economic development

2 by a measure known as energy intensity, which is assessed by calculating the units of energy used per unit of GDP. The fewer the units of energy, the more efficiently a country is using its energy supply. In general, energy intensity values decrease with economic development.

Exam tip

There may be questions in the exam where you should show that you are aware of this distinction between primary and secondary energy.

Energy mix is the combination of different energy sources available to meet a country's total energy demand.

Primary energy is any form of energy found in nature that has not been subject to any conversion or transformation. Primary energy can be renewable (water and wind power) or non-renewable (coal, oil and gas).

Secondary energy refers to the more convenient forms of energy, such as electricity, which are derived from the transformation or conversion of primary energy sources.

Knowledge check 89

Define the term 'recyclable energy'.

Knowledge check 90

What are the advantages of converting primary energy sources into electricity?

Figure 76 shows some of the factors affecting per capita energy consumption. The upper five factors in the diagram have already been touched on and need little further explanation, but the lower three perhaps do:

- Public perceptions of, or attitudes towards, energy differ. Many consumers are worried by nuclear powers safety risks, but also find wind turbines ugly. Coal is widely perceived as 'dirty' while natural gas (also a fossil fuel) is seen as clean or even 'green' by comparison.
- Climate is a significant factor. Very high levels of consumption in North America, the Middle East and Australia partly reflect the widespread use of air-conditioning in homes and businesses in the summer, and heating in the winter in countries such as Canada.
- Environmental priorities of governments. For some, the energy policy will be one of taking the cheapest route to meeting the nation's energy needs, regardless of the environmental costs. Others will seek to increase their reliance on renewable sources of energy, while still others will have in place policies that raise energy efficiency and energy saving.

Knowledge check 91

How does technology affect per capita energy consumption?

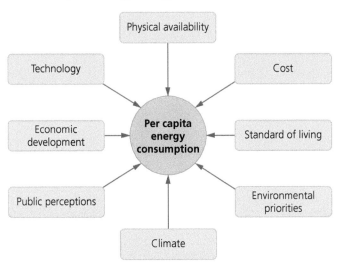

Figure 76 Some factors affecting per capita energy consumption

Energy portraits of France versus the USA

The USA and France rank second and tenth respectively in the league table of energy consumers, but total energy consumption in France is about one-tenth that of the USA. The difference is largely explained by differences in population — 330 million in the USA compared with 65 million in France. In per-person terms, the USA is the 11th largest energy consumer and France the 31st.

Figure 77 (p. 130) compares the USA and France in terms of their sources of primary energy. In the USA, over three-quarters of the energy comes from fossil fuels. The French energy mix is very different, with half its energy coming from fossil fuels and around 40% coming from nuclear energy. In terms of energy security, France is much less well placed than the USA, if only because nearly half of its primary energy is imported. The USA is much more self-sufficient with large domestic fossil fuel resources — but has significantly higher carbon emissions per person.

Knowledge check 92

Why is the USA less dependent on nuclear energy than France?

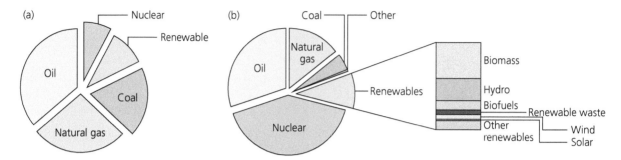

Figure 77 Energy consumption by primary source: (a) the USA and (b) France

Energy players

Meeting the demand for energy involves **energy pathways** from producer to consumer. At both ends of such pathways there are influential players (organisations, groups or individuals) with a particular involvement in the energy business. At the supply end, there are energy companies and the governments of energy-producing countries. There are governments at the demand end also, as well as a range of consumers from industrial to domestic. Along the pathways there are companies responsible for the movement and processing of energy. For more on energy pathways see p. 132.

An **energy pathway** is the route taken by any form of energy from its source to its point of consumption. The routes involve different forms of transport, such as tanker ships, pipelines and electricity transmission grids.

Synoptic link

(P): Players in the energy business have very different motives, from profit (TNCs, OPEC) to national security (governments) and low-cost, reliable energy (consumers), which can lead to tensions between players.

Exam tip

Be sure you understand why OPEC is such a powerful player in the global energy business.

Table 57 sets out the major players in the world of energy. Of particular importance here is OPEC.

Table 57 Major players in the world of energy

Players	Role
Transnational corporations (TNCs)	The big names in the oil and gas business include Gazprom, ExxonMobil, PetroChina and Royal Dutch Shell. Nearly half of the top 20 companies are state-owned (all or in part) and therefore very much under government control (reflecting the importance of energy in national security). Most are involved in a range of operations: exploring, extracting, transporting, refining and producing petrochemicals
Organization of the Petroleum Exporting Countries (OPEC)	OPEC has 13 member countries which between them own around two-thirds of the world's oil reserves. Because of this, it is in a position to partly control the amount of oil entering the global market, as well as influence its price. OPEC has been accused of holding back production in order to drive up oil prices
Energy companies	Important here are the companies that convert primary energy (oil, gas, water and nuclear) into electricity and then distribute it, e.g. EDF and E.ON in the UK. Most companies are involved in the distribution of both gas and electricity. They have considerable influence when it comes to setting consumer prices and tariffs
Consumers	An all-embracing term, but probably the most influential consumers are transport, industry and domestic users. Consumers are largely passive players when it comes to fixing energy prices
Governments	They can play a number of different roles — they are the guardians of national energy security and can influence the sourcing of energy for geopolitical reasons

Reliance on fossil fuels

Despite growing global concern about increasing carbon emissions and their contribution to climate change, the world continues to rely on fossil fuels (coal, oil and natural gas) for over 80% of its energy needs.

Mismatch between fossil fuel supply and demand

A fundamental feature of energy resources is that the geographical distributions of fossil fuel supply and demand are not the same. This important point is shown in Tables 58, 59 and 60, which set out the top ten producers and consumers of each of the three main fossil fuels.

Table 58 The world's leading coal producers and consumers, 2019

Coal production		Coal consumption	
Country	Production (m tonnes)	Country	Consumption (m tonnes)
China	3846	China	2866
India	756	India	585
USA	640	USA	397
Indonesia	610	Germany	183
Australia	507	Japan	163
Russia	440	Russia	162
South Africa	254	South Korea	150
Germany	134	South Africa	130
Kazakhstan	115	Poland	122
Poland	112	Indonesia	115

Table 59 The world's leading crude oil producers and consumers, 2019

Oil production		Oil consumption	
Country	Production (m tonnes)	Country	Consumption (m tonnes)
USA	747	USA	842
Russia	568	China	650
Saudi Arabia	557	India	242
Canada	275	Japan	174
Iraq	234	Saudi Arabia	158
China	191	Russia	151
UAE	180	South Korea	120
Iran	161	Brazil	107
Brazil	151	Germany	106
Kuwait	144	Canada	102

Table 60 The world's leading natural gas producers and consumers, 2019

Gas production		Gas consumption	
Country	(billion m³)	Country	(billion m³)
USA	921	USA	847
Russia	679	Russia	444
Iran	244	China	307
Qatar	178	Iran	224
China	178	Canada	120
Canada	173	Saudi Arabia	114
Australia	153	Japan	108
Norway	114	Mexico	91
Saudi Arabia	113	Germany	89
Algeria	86	UK	79

Energy pathways

It should now be clear from Tables 58, 59 and 60 that there are basic mismatches between the distributions of production and consumption of all three main fossil fuels. Japan, Germany and South Korea are major consumers but not producers. These mismatches are resolved by the creation of energy pathways (see p. 130) that allow transfers to take place between producers and consumers. The main fossil fuel pathways are:

- Oil is transported by tanker ships from the Middle East to Europe, Asia and North America and also by pipelines and road/rail freight.
- Gas is mainly moved by pipeline, especially from Russia to Europe, but also by liquefied natural gas (LNG) ship tankers from the Middle East.
- Coal is mainly transported by bulk-carrier ship and rail: its lower energy density and high weight make transport costs high.

Russian gas to Europe

Energy pathways are a key aspect of energy security but can be prone to disruption, especially as conventional fossil fuels have to be moved over long distances from sources to markets. Table 60 shows that Russia is the second largest producer of gas. Most of its gas exports go to European countries. Russian gas is delivered to Europe mainly through five pipelines.

Geopolitically significant is the fact that three of those pipelines cross Ukraine, a country from which Russia annexed the Crimea in 2014. It also now occupies parts of eastern Ukraine. Clearly, Ukraine might be in a position of strength here. It could increase its charges for allowing Russian gas to pass through it. Gas supplies were disrupted in 2006 and 2009 due to disputes between Russia and Ukraine. Germany is partly funding the Nord-Stream 2 gas pipeline (opening 2021) through the Baltic to bypass the Ukraine problem, and other European countries increasingly rely on imported LNG from Qatar.

Given the history of strained political relations between Russia and Western Europe, it would appear strategically unwise for EU countries to become heavily reliant on Russian gas and oil (Figure 78). The UK still obtains most of its imported gas from Norway, though it has recently substantially increased its imports of gas from Qatar in order to offset the declining output from its North Sea gas fields.

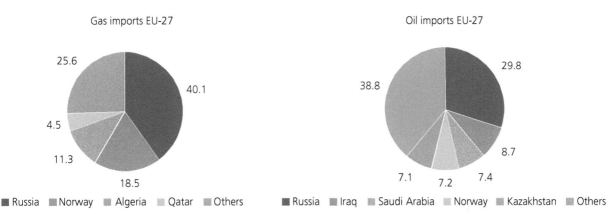

In 2018, 60% of the EU-27 energy needs were met by imports.
In Germany, import dependency was 64%

Figure 78 EU-27 dependency on oil and gas imports, 2018

Source: Eurostat

Unconventional fossil fuels

Despite the need to move the global energy budget towards renewable energy sources, much exploration work is still going on in the search for new oil and gas fields. At the same time, attention has turned towards what are called 'unconventional fossil fuels'. There are four: tar sands, oil shale, shale gas and deepwater oil. Canada is leading the way with the first of these, the USA with the second and third, and Brazil with the last (Table 61).

Exam tip

The gas pipelines from Russia to Western Europe illustrate well the risks of transboundary energy pathways.

Knowledge check 93

Why is the output from the UK's gas fields in the North Sea declining?

Table 61 Unconventional sources of fossil fuel

Resource	Nature	Extraction
Tar sands	A mixture of clay, sand, water and bitumen (a heavy, viscous oil)	Tar sands have to be mined and then injected with steam to make the tar less viscous so that it can be pumped out
Oil shale	Oil-bearing rocks that are permeable enough to allow the oil to be pumped out directly	Either mined, or shale is ignited so that the light oil fractions can be pumped out
Shale gas	Natural gas that is trapped in fine-grained sedimentary rocks	Fracking: pumping in water and chemicals forces out the gas
Deepwater oil	Oil and gas that are found well offshore and at considerable oceanic depths	Drilling takes place from ocean rigs; already under way in the Gulf of Mexico and off Brazil

It is important to note that exploitation of these unconventional sources has a downside:

- They are all fossil fuels, so their use will continue to threaten the carbon cycle and contribute to global warming.
- Extraction is costly and requires a high input of complex technology, energy and water.
- They all threaten environmental damage, from the scars of opencast mines and land subsidence to the pollution of groundwater and oil spills. Certainly, the resilience of fragile environments will be sorely tested.

The last of these bullet points clearly signals social costs in the form of degraded residential environments and disrupted communities. But are there any social benefits to be offset against these costs? Possibly energy companies might invest in improving the local social infrastructure as a sweetener?

Players in the harnessing of unconventional fossil fuels have conflicting views.

- Exploration companies: they have a key role to play in discovering and developing reserves. However, they will be keen to see a good financial return on their exploratory work and perhaps willing to take risks with the environment in order to achieve this.
- Leading oil and gas TNCs: they are anxious that their investments in conventional fossil fuels are not threatened by competition from these unconventional sources.
- Governments: some will see domestic sources of these fuels as offering a higher level of energy security, while others might wish to avoid any political fallout led by environmental groups and affected communities.
- Environmental groups: these are well organised and vocal in pointing out risks and potential damage to the environment. Clearly, they favour renewable energy sources.
- Affected communities: these may well be divided between those that support exploitation of the sources (on the grounds of providing jobs and generating local income) and those that see the peace and quiet of their home areas and environmental quality being threatened.

> **Exam tip**
>
> It is unlikely that the unconventional sources of fossil fuel will ever challenge the conventional ones.

> **Exam tip**
>
> Remember that particular types of player, for example governments or affected communities, do not always share the same attitudes.

> **Synoptic link**
>
> (P): The views of different players involved in unconventional fossil fuels are almost impossible to reconcile because of the serious environmental consequences versus the profit to be gained.

Alternatives to fossil fuels

Renewable and recyclable energy

The global drive to reduce carbon dioxide emissions must involve increasing reliance on alternative sources of 'clean' energy, so decoupling economic growth from dependence on fossil fuels. Basically, this means widening the energy mix to include substantial inputs from both renewable and recyclable energy sources.

The main sources of renewable energy today are hydro, wind, solar (mainly via photovoltaic cells), geothermal and tidal. The contribution made by these sources to national energy budgets varies from country to country. It is a simple fact of geography that not all countries have renewable energy resources to exploit. For example, not all countries have coasts, strongly flowing rivers or climates with either long sunshine hours or persistently strong winds. Partly because of this there are very few countries where renewables might completely replace all the energy currently derived from fossil fuels. Other factors to consider include:

- the relative financial costs of using non-renewable and renewable energy sources. When oil and gas prices are low, renewables become a more expensive option, but costs of wind and solar energy have declined rapidly in the last 10 years
- the harnessing of renewables is not without environmental costs. Think of the drowning of river valleys to create HEP reservoirs, or the large areas of land and the offshore zone that will be covered by the required number of solar and wind farms
- while the majority of people believe that we should make greater use of renewable sources, many go off the idea when it is proposed to construct a wind or solar farm close to where they live.

Some countries have used nuclear energy to generate their electricity supply in a reasonably carbon-free manner. France gets 70% of its electricity from nuclear power, useful in a country with limited domestic fossil fuel resources. Nuclear waste can be reprocessed and reused, making it into a recyclable energy source.

The use of nuclear energy does have downsides, which include:

- risks to do with safety (accidents) and security (terrorism)
- the disposal of high-level radio-active waste with an incredibly long decay life
- the technology involved is complex and therefore its use is usually only an option for developed countries
- public attitudes to nuclear power are often hostile, such that it has been abandoned in Germany and scaled back in many countries
- although the operational costs are low, the costs of constructing and decommissioning power stations are high.

Photovoltaic cells are now the most widely used method for generating electric power using solar cells to convert energy from the sun into an electric current that can be used to power equipment or recharge batteries.

Knowledge check 94

Why do people object to wind and solar farms being built near their homes?

Exam tip

Be prepared to cite an example of a nuclear power station incident, such as that at Chernobyl (Ukraine) and Fukushima (Japan) — the latter caused by an earthquake and its tsunami.

UK's changing energy mix

Figure 79 (p. 136) shows that when it comes to electricity generation in the UK, there has been a complete shift away from coal. The gap in the energy mix has been largely filled by an increased use of wind, solar and bioenergy. The UK has rapidly moved to lower-carbon renewables and decreased fossil fuel consumption. At the same time falling total electricity generation suggests more efficient use and lower use per person. However, UK industry still relies heavily on natural gas, and oil is used by transport (petrol and diesel), and this is only slowly changing as the use of electric vehicles increases.

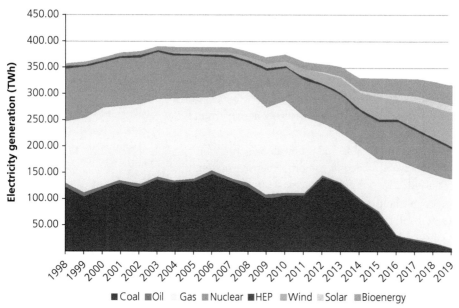

Figure 79 The UK's changing electricity generation, 1998–2019

Biofuels

Of all the energy sources used by humans, fuelwood perhaps has the longest history. However, while fuelwood remains important in the energy mix in some parts of the world, **biomass** has recently come to prominence with the commercial use of a number of relatively new biofuels.

The growing of **biofuel** crops is being increasingly recognised as one way of reducing both the burning of fossil fuels and carbon dioxide emissions. The most widely grown biofuel crops include wheat, corn, grasses, soy beans and sugar cane. In the UK, the two main crops are oilseed rape and sugar beet. Most of these two crops are converted into ethanol or biodiesel, which are mainly used as a vehicle fuel.

But while the use of these organic fuels is to be commended on the grounds that they are 'green' energy sources, there is a serious downside to them. Each hectare of farmland used to grow energy crops means a hectare less for growing much-needed food in an increasingly hungry world. Added to this, there is still some uncertainty over how carbon-neutral biofuel crops really are.

Biofuels in Brazil

Since the 1970s, Brazil has taken steps to diversify its energy mix and improve its energy security. This drive has been spearheaded by developing the country's considerable hydro power resources. More recently, it has added biofuels to its energy portfolio. Although less than 5% of Brazil's energy comes from renewable energy sources, 90% of new passenger vehicles sold in the country have flex-fuel engines that work using any combination of petrol and ethanol. This has led to a significant reduction in the country's carbon emissions.

Large areas of central southern Brazil are now set aside for the cultivation of sugar cane and the subsequent production of ethanol. The result has been the displacement of other types of agriculture, particularly cattle rearing. The need to find replacement

Knowledge check 95

What change in the UK economy has helped to reduce the total consumption of energy?

Biomass is organic matter used as a fuel, for example in power stations for the generation of electricity.

Biofuel is derived directly from organic matter, such as agricultural crops, forestry or fishery products, and various forms of commercial and domestic waste. **Primary biofuels** include fuelwood, wood chips and pellets, as well as other organic matter, that are used in unprocessed form, primarily for heating, cooking and electricity generation. **Secondary biofuels** are derived from the processing of biomass and include liquid biofuels, such as ethanol and biodiesel, which can be used in motor vehicles and some industrial processes.

pastures has had a serious knock-on effect. It has resulted in the large-scale clearance of tropical rainforest in the Amazon basin. Ironically, this deforestation nullifies the reduction in carbon dioxide emissions gained from the increasing use of ethanol.

Radical technologies to reduce carbon emissions

Finally, it is necessary to consider two rather more radical technologies. These promise to reduce carbon emissions while the world continues to meet the ever-rising demand for energy.

1 Carbon capture and storage involves 'capturing' the carbon dioxide released by the burning of fossil fuel and burying it deep underground. Unfortunately, it is an expensive process because of the complex technology involved. There is also some uncertainty over whether the stored carbon will stay trapped underground and that it will not slowly leak to the surface and into the atmosphere.

Since it is widely accepted that fossil fuels will continue to provide most of the world's primary energy, development of the carbon capture and storage technology must be given a high priority, as must also the slightly different technology that 'scrubs' some of the carbon dioxide out of exhausts produced by the burning of fossil fuels.

2 Hydrogen fuel cells combine hydrogen and oxygen to produce electricity, heat and water. They will produce electricity as long as hydrogen is supplied; they will never lose their charge. They are a promising technology for use as a source of heat and electricity for buildings, and as a power source for electric vehicles.

The challenge with this technology is finding a cheap and easy source of hydrogen. Although it is a simple and abundant chemical element, it does not occur naturally as a gas. It is always combined with other elements, for example with oxygen in water. It may be possible to produce hydrogen using renewable electricity from wind or solar power in the future.

A world free from the need to burn fossil fuels for energy is highly improbable. However, a world deriving much of its energy from renewable and recyclable sources, and making full use of the hydrogen fuel cell, does promise much less disturbance of the carbon cycle, its stores and fluxes.

How are the carbon and water cycles linked to the global climate system?

- The growing demand for food, fuel and other resources is threatening biological carbon stores, the water cycle and soil health.
- The degradation of both carbon and water cycles is having an adverse impact on human wellbeing.
- Global warming and climate change require co-ordinated international efforts to reduce carbon emissions and to devise effective adaptive and mitigating strategies.

Exam tip

Be sure that you know the names of some biofuel crops and that they are mainly converted into fuels for motor vehicles.

Carbon and water cycles threatened by human activity

This section focuses on three ways in which the biological carbon cycle is being disrupted by human activities. The first of these is directly by resource exploitation and associated land-use changes. The other two ways are the indirect consequences of climate change and the enhanced greenhouse effect.

Growing resource demands

The burning of fossil fuels is not the only human activity that is disturbing the biological carbon and hydrological cycles. There are others related to the growing global demand for food, fuel and other resources, all of which are the outcome of continuing global population growth and economic development.

■ Deforestation: the clearance of forests both for their timber and for the land they occupy. In the latter case, the land is mainly cultivated to provide grazing for livestock or to produce cash crops. However, as Figure 80 shows, it is not all bad news in that there is both reforestation and afforestation under way in temperate latitudes. This is helping to offset the loss of tropical rainforest 'services', but in the case of afforestation much is taking place on what was agricultural land.

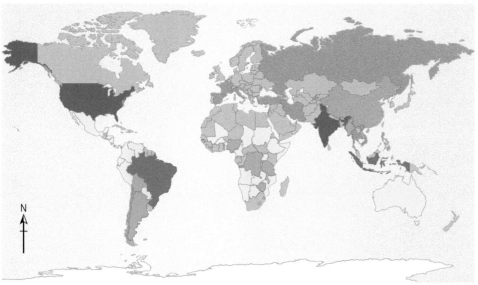

Key

1,000 ha

Net gain	Net loss	Small change (gain or loss)
▨ 50–250	■ >500	▫ <50
■ 250–500	▨ 500–250	
■ >500	▯ 250–50	☐ No data

Figure 80 Annual change in forest areas, 1990–2015

■ Grassland conversion: temperate and tropical grasslands have also become heavily exploited by agriculture. Both grassland types have suffered as a result of over-exploitation. The simple act of ploughing leads to an immediate loss of both carbon dioxide and moisture, as well as a change in runoff characteristics.

■ Urbanisation: no land-use conversion is greater than that associated with urbanisation. Much space has already been taken over and many ecosystems destroyed by the insatiable demand for space needed to accommodate a rapidly rising urban population and the widening range of economic activities. Of all forms of development, none is having a more disruptive impact on the carbon and water cycles than urbanisation. Towns and cities are focal points of both GHG emissions and intense water demand.

Clearly, these changes vary individually from place to place and as a consequence so does their overall impact on carbon stores, soil health and the water cycle. In some locations, the impact is considerable; in others minimal if at all.

Ocean acidification

Ocean acidification is very much the outcome of climate change related to the burning of fossil fuels. It represents a serious disturbance of the biological carbon cycle. The acidification is caused by the oceans being important **carbon sinks** in the carbon cycle. Up to the early 19th century, the average oceanic pH was 8.2. By 2019 it had fallen to 8.1. This may seem a minuscule change, but the mean values disguise the fact that there has been a larger fall in the pH of surface waters (Figure 81). Coral reefs, an important component of ocean life, stop growing when the pH is less than 7.8.

> **Ocean acidification** involves a decrease in the pH (alkalinity) of the oceans caused by the uptake of carbon dioxide from the atmosphere.
>
> A **carbon sink** is any natural environment (a forest, wetland or ocean) that is capable of absorbing more carbon dioxide from the atmosphere than it releases to the atmosphere. The carbon sink function is the precursor to a particular environment becoming a carbon store.

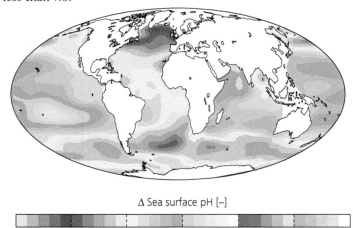

Δ Sea surface pH [−]

−0.12   −0.1   −0.08   −0.06   −0.04   −0.02   0

Figure 81 Estimated change in sea water pH between the 1700s and 1990s

So the situation is now approaching the point that there is a real risk of some marine ecosystems and their goods and services passing the critical threshold of permanent damage. In the case of coral reefs, they are also being threatened by the rise in surface water temperatures. The widespread bleaching of coral in the Great Barrier Reef of Australia is a clear indication that the threat has become a reality.

> **Knowledge check 96**
>
> What is the difference between a carbon sink and a carbon store?

Health of forests

Like ocean acidification, the declining health of forests is also the outcome of the enhanced greenhouse effect and consequent climate change. The health of the world's forests as carbon stores is being challenged in three ways:

1 by deforestation
2 by the poleward shift of climatic belts
3 by increasing drought.

The three are related in that the first and second are factors encouraging the third.

Amazon droughts

This is well illustrated by the Amazon rainforest, which acts as a giant climate regulator. Every day, it pumps 20 billion tonnes of water into the atmosphere. This is 3 billion tonnes more than the River Amazon discharges into the Atlantic Ocean. The forest's uniform humidity lowers atmospheric pressure, allowing moisture from the Atlantic to reach almost across the continent. However, since 1990, a cycle of extreme drought and flooding has been observed. Droughts in 2005, 2010 and 2016 greatly degraded much of the forest already stressed by prolonged and large-scale deforestation.

In short, the diminishing health of the tropical rainforest means that it is:

- declining as a carbon store
- sequestering less carbon dioxide from the atmosphere, thereby exacerbating the greenhouse effect
- playing a diminished role in the hydrological cycle.

Implications for human wellbeing

First, the point should be made that while human activities are largely responsible for the climate change resulting from the enhanced greenhouse effect, the consequent disruption of the carbon and water cycles is having a negative rebound effect on human wellbeing.

Impacts of forest loss

It is now widely understood that the impacts of deforestation are global in scale and not just confined to deforested areas. We now understand the value of forests, for example in:

- sequestering carbon dioxide from the atmosphere
- storing carbon
- transferring moisture from the soil back into the atmosphere by evapotranspiration.

It looks as if the Kuznets curve (Figure 82) is correct in suggesting that as they reach higher levels of development and wealth, societies approach a tipping point when the costs of resource exploitation become fully realised and are set against the benefits of resources conservation and protection.

UK forests

After centuries of deforestation, the forest cover of the UK had been reduced from an estimated original figure of 80% to less than 10% by the end of the 19th century. The Forestry Commission was set up in 1919 to remedy the country's shortage of timber. It started to plant fast-growing exotic confers, such as Sitka spruce, on the moors of Wales, in the Scottish Highlands and in the English Lake District and Pennines. Today 13% of the UK's land surface is now forested. In recent years, the cultivation of exotic conifers has given way to the planting of indigenous species. Today, there is much less emphasis on the commercial production of timber and more on the environmental benefits of restoring a forest cover close to the original.

So as more and more countries put the brake on deforestation and instead begin programmes of reforestation (as in the taiga), so forest loss eventually begins to have what might be seen as a positive impact.

Exam tip

Be clear that the health of tropical rainforests is much more threatened by deforestation and climate change than that, say, of the boreal forests.

Knowledge check 97

Why is the health of the boreal forests less threatened than that of the tropical rainforest?

Knowledge check 98

What is a tipping point?

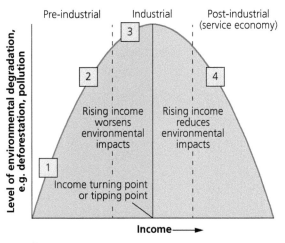

Key
1. UK pre-Industrial Revolution, remote Amazonia today, Indonesia pre-1970s
2. Indonesia today, China in the 20th century
3. China today
4. UK today

Figure 82 The environmental Kuznets curve

Unfortunately, the same cannot be said for the next two changes. Their negative impacts are beginning to be understood, but as yet little remedial action is being undertaken. But will it really take, as the Kuznets curve suggests, further increases in wealth before the tide turns from exploitation to conservation?

> **Synoptic link**
>
> (A): Players have very different attitudes towards sustainability and on environmental issues. Consumers have become more aware of environmental issues in the last 20 years, but less so when acting in the interest of the environment has an economic cost.

Impacts of rising temperatures

The rising temperatures resulting from GHG emissions are increasing both evaporation rates and the amount of water vapour. This, in turn, is impacting on:

- precipitation patterns
- river regimes
- drainage basin stores
- the cryosphere.

The Arctic

The Arctic plays an important role in global climate as its sea ice regulates evaporation and precipitation. What has happened here over the last few decades serves as a warning to the rest of the planet:

- Temperatures have risen 2.5°C since the 1970s, twice as fast as the global average.
- Arctic sea ice has declined by 40% since 1978 — the Northwest Passage is now open to summer navigation.

- There has been much melting of the permafrost.
- Carbon uptake by terrestrial plants is increasing because of a lengthening growing season.
- There has been a loss of albedo as the ice that once covered the land surface gives way to tundra and tundra gives way to taiga. Sunlight that was previously reflected back into space by the white surface is now being increasingly absorbed by the ever-darkening land surface. In other words, it is encouraging further global warming.

In terms of human wellbeing, there have been both pluses and minuses. The warming climate is opening up previously ice-bound wilderness areas to tourism. The exploitation of mineral resources, particularly Arctic oil and gas, is becoming more feasible. However, climate warming is disrupting and destroying traditional ways of life, for example of the fishing and hunting Inuits of North America and the Sami reindeer herders of northern Eurasia.

Knowledge check 99

What is albedo?

Synoptic link

(F): Although scientific understanding of global warming is increasing, there is still much uncertainty about the future. This means there is a wide range of projections for warming, precipitation and sea level and uncertainty increases the further into the future we try to project.

Exam tip

Learn some key facts and figures about a case study, such as that of the Arctic, as these add weight to your answers.

Impacts of declining ocean health

The decline in ocean health caused by acidification and bleaching is resulting in changes to marine food webs. In particular, fish and crustacean stocks are both declining and changing their distributions. Such changes are being particularly felt by developing countries.

- The FAO estimates that fishing supports 500 million people, 90% of whom live in developing countries.
- Millions of fishing families depend on seafood for income as well as food.
- Seafood is also the dietary preference of some wealthier countries, notably Iceland and Japan.
- Aquaculture is on the rise, but its productivity is also being affected by declining pH values and rising temperatures.

Tourism is another activity under threat, particularly in those countries, for example in the Caribbean, where coral reefs, now showing signs of degradation, have traditionally attracted scuba-diving tourists. The rising sea level is yet another consequence of climate change that threatens the very survival of tourism and its coastal infrastructure, as for example in the Maldives and the Seychelles. The costs of strengthening coastal defences can often exceed the financial resources of poorer coastal countries.

Exam tip

It is not all doom and gloom. There is growing evidence that some marine life is already shifting polewards to compensate for the warming seas.

Responses to the risk of further global warming

An uncertain future

There is much uncertainty about the future (Figure 83), which raises many questions, particularly:

- the level of GHG emissions — will they continue to rise and how fast?
- the degree of concentration in the atmosphere — is there a limited capacity?
- the resilience of other carbon sinks — what are their capacities and could they store more?
- the degree of climate warming — how much warmer?
- feedback mechanisms such as the release of carbon from peatlands and thawing permafrost — what volumes of carbon are likely to be released?
- the rate of population growth — when, if ever, will it level off?
- the nature and rate of economic growth — will it always be so carbon-based?
- the harnessing of alternative energy sources — will fossil fuels be completely replaced?
- the possible passing of tipping points relating to such aspects as forest dieback and irreversible alterations to the thermohaline circulation — will disaster be sure to follow?

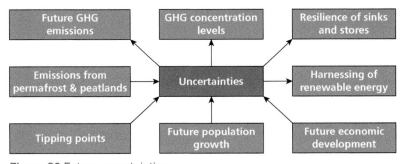

Figure 83 Future uncertainties

Figure 83 underlines the point made earlier, that any forecasting of global futures should be undertaken with the greatest caution. There is still so much that is unknown.

The inevitable and critical question is this: how should we react to this very real threat of further global warming? There are two different, but not incompatible, courses of action:

1. Adaptation: changing our ways of living in such a manner that we are able to cope with most, if not all, of the outcomes of global warming.

2. Mitigation: reducing or preventing GHG emissions by devising new technologies and adopting low-carbon energies (renewables and recyclables).

In the present context of global warming, adaptation is, in a sense, make do and mend — that is, living with the problem, not solving it. Mitigation, on the other hand, seeks to tackle the root cause of the problem.

Exam tip

It is important to be certain about the difference between adaptation and mitigation.

Adaptation strategies for a changed climate

Table 62 sets out an evaluation of five adaptation strategies or courses of action:

1 Water conservation and management.

2 Resilient agricultural systems.

3 Land-use planning.

4 Flood-risk management.

5 Solar radiation management.

Four of the strategies involve a mix of soft- and hard-engineering actions. Some of those actions are low in technology and upfront costs and so, in theory, are possible options for developing countries. A change in traditional practices and customs is often required here. However, there are also actions requiring high inputs of capital and technology that only developed countries can contemplate. The whole of the solar radiation management strategy clearly falls into this category.

Table 62 Costs and benefits of adaptation

Adaptation strategies	Benefits	Costs and risks
Water conservation and management	Fewer resources used, less groundwater abstraction Attitudinal change operates on a long-term basis: use more grey water (recycled water)	Efficiency and conservation cannot match increased demands for water Changing cultural habits of a large water footprint need promotion and enforcement by governments, e.g. smart meters
Resilient agricultural systems	Higher-tech, drought-tolerant species help resistance to climate change and increase in diseases Low-tech measures and better practices generate healthier soils and may help CO_2 sequestration and water storage: selective irrigation, mulching, cover crops, crop rotation, reduced ploughing, agroforestry More 'indoor' intensive farming	More expensive technology, seeds and breeds unavailable to poor subsistence farmers without aid High energy costs from indoor and intensive farming Genetic modification is still debated but increasingly used to create resistant strains, e.g. rice and soya Growing food insecurity in many places adds pressure to find 'quick fixes'
Land-use planning	Soft management: land-use zoning, building restrictions in vulnerable flood plains and low-lying coasts Enforcing strict runoff controls and soakaways	Public antipathy Abandoning high-risk areas and land-use resettling is often unfeasible, as in megacities such as Dhaka, Bangladesh or Tokyo-Yokohama, Japan A political 'hot potato' Needs strong governance, enforcement and compensation

Adaptation strategies	Benefits	Costs and risks
Flood-risk management	Hard management traditionally used: localised flood defences, river dredging Simple changes can reduce flood risk, e.g. permeable tarmac Reduced deforestation and more afforestation upstream to absorb water and reduce downstream flood risk	Debate over funding sources, especially in times of economic austerity Landowners may demand compensation for afforestation or 'sacrificial land' kept for flooding Constant maintenance is needed in hard management, e.g. dredging; lapses of management can increase risk Ingrained culture of 'techno-centric fixes': a disbelief that technology cannot overcome natural processes
Solar radiation management (SRM)	Geoengineering involves ideas and plans to deliberately intervene in the climate system to counteract global warming The proposal is to use orbiting satellites to reflect some inward radiation back into space, rather like a giant sunshade It could cool the Earth within months and be relatively cheap compared with mitigation	Untried and untested Would reduce but not eliminate the worst effects of GHGs; for example, it would not alter acidification Involves tinkering with a very complex system, which might have unintended consequences or externalities Would need to continue geoengineering for decades or centuries as there would be a rapid adjustment in the climate system if SRM stopped suddenly

Knowledge check 100

For one of the adaptation strategies in Table 62, give examples of soft- and hard-engineering actions.

Exam tip

Be sure that you are aware of at least two of the strategies in Table 62 in terms of their benefits and costs, as well as their feasibility in different parts of the world.

Mitigation and rebalancing the carbon cycle

The long-term solution to the global warming crisis lies in rebalancing the carbon cycle, particularly reducing the concentration of GHGs in the atmosphere. This requires taking actions that fall under the heading of mitigation. Table 63 sets out seven possible mitigations. None is straightforward, except possibly afforestation. Successful implementation requires a society to change the way it thinks and acts. Some mitigation has a high technological tariff.

Table 63 Mitigation methods applied to the UK

Method	Policy
Legal framework	Since 2019, the UK has had a target of net zero emissions by 2050, with carbon budgets setting out the pathway to this goal
Energy sector	By 2024–2025, coal will no longer be used to generate electricity Low-carbon electricity producers, such as wind and solar, are guaranteed a minimum price for electricity (a subsidy) to encourage investment
Taxation	Tax on petrol and diesel sales, and annual car tax linked to carbon emissions per km, are at least in part 'green taxes'
Transport	In 2030 it is planned to phase out sales on new petrol and diesel cars in favour of electric and hybrid vehicles. This should reduce fossil fuel use, as long as electricity generation for the national grid continues to move towards renewable energy
Efficiency	Cars, homes and electrical appliances all have energy ratings which indicate their energy efficiency. This means consumers can make informed choices
Carbon capture	Carbon capture and storage (CCS) has not yet fulfilled its promise, but the UK government has invested in a number of small pilot schemes
Homes	Several policies such as the 'Green Deal' and 'Green Homes Grant' have been used to encourage householders to fit insulation, double glazing and other energy-saving technology by providing financial grants. These schemes have tended to be less successful than planned

Knowledge check 101

Apart from renewable energy, which of the methods in Table 63 do you think is the most feasible?

There are two other important points to be made about mitigation and, to some extent, about adaptation too.

1 The first is to recognise that there is a range of possible human intervention options and targets that runs from 'business as usual' (but perhaps making some adaptations) to 'aggressive mitigation' (Figure 84). RCPs (recommended concentration pathways) are four different concentrations of GHGs in the atmosphere identified by the IPCC (Intergovernmental Panel on Climate Change). The sobering message is that even with strong mitigation measures, there is no guarantee that even if emissions are halved by 2080 the mean global temperature will not rise by more than 2°C.

> **Exam tip**
>
> Remember that the four RCPs refer to the concentration of GHGs in the atmosphere, not to the rate of emission.

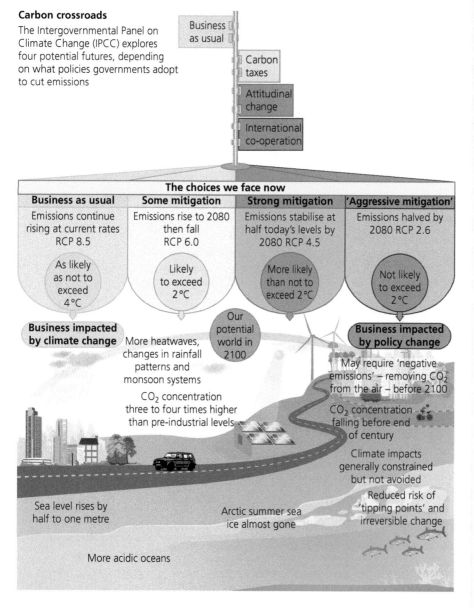

Figure 84 The carbon crossroads

2 The second is that if mitigation, at whatever pitch, is to have any chance of success, it not only requires concerted actions at a national level but, more critically, it requires effective international agreements. Global warming is a global problem requiring global action.

The latter point was first accepted in 1997 by the Kyoto Protocol, an international agreement which aimed to cut GHG emissions by 5% by 2012. Since then, the reduction targets have been revised upwards and emissions have been reduced. It remains to be seen whether enough is being done or whether the global mitigation strategy should be made even more aggressive.

It has to be said that not every country has been enthusiastic about signing up to the succession of agreements tabled since 1997. The most recent of these, the Paris Agreement of 2015, aims to keep the rise in the global temperature to less than 2°C above its pre-industrial level. The Agreement now has 190 national signatures. Among the more reluctant signatories are the three largest producers of GHGs: China, India and the USA (withdrew in 2017, rejoined in 2021).

Knowledge check 102

Why is international agreement the key to any successful mitigation of global warming?

Exam tip

Learn the locations and dates of important IPCC meetings: Kyoto (1997), Copenhagen (2009) and Paris (2015).

Synoptic link

(A): Attitudes to global climate and environmental threats are changing, but slowly and unevenly. Some governments (EU) commit to mitigation but others are more reluctant (USA, Australia) and the same is true of TNCs and consumers. The overall 'direction of travel' is towards actions to mitigate global warming, but progress is likely to be slow.

Summary

- Most global carbon is locked in terrestrial stores (rocks) as part of a long-term geological cycle.
- Biological processes sequester carbon on land and in the oceans, but on a short time cycle.
- A balanced carbon cycle is important in sustaining other Earth systems, but it is being increasingly disrupted by human activities.
- Energy security is a key goal for all countries, but it needs to be achieved with much less dependence on fossil fuels.
- Economic development still relies heavily on fossil fuels as a source of energy.
- Maintaining energy pathways is a vital aspect of energy security.
- There are energy alternatives to fossil fuels, but each has its costs and potential benefits.
- Both the biological carbon and water cycles are threatened by human activities.
- The threat comes largely from the growing demand for food, fuel and other resources.
- The degradation of carbon and water cycles threatens human wellbeing, particularly in developing countries.
- Further global warming risks releasing more carbon and as such needs responses from different players at different scales.

Questions and Answers

■ Assessment overview

In this section of the book questions on each of the A-level content areas are given. The style of questions used in the examination papers has been replicated, with a mixture of short answer questions, data response questions and extended writing questions.

All questions that carry a large number of marks require candidates to consider connections between the subject matter and to demonstrate deeper understanding in order to access the highest marks.

The breakdown of the questions per topic is:
- SECTION A: Tectonic processes and hazards (16 marks)
  Typical question sequence: 4- and 12-mark questions
- SECTION B: Landscape systems, processes and change (**either** Glaciated landscapes and change **or** Coastal landscapes and change) (40 marks)
  Typical question sequence: 6-, 6-, 8- and 20-mark questions
- SECTION C: Physical systems and sustainability (Water cycle and Carbon cycle) (49 marks)
  Typical question sequence: 3-, 6-, 8-, 12- and 20-mark questions

The paper takes 2 hours and 15 minutes and is worth a total of 105 marks, making up 30% of the A-level qualification.

In this section of the book, each of the content areas is structured as follows:
- sample questions in the style of the examination
- levels-based mark schemes for extended questions (6 marks and over) in the style of the examination
- one sample student answer per question — at the upper level
- examiner commentary on each of the above.

Carefully study the descriptions given after each question to understand the requirements necessary to achieve a high mark. You should also read the commentary with the mark schemes to understand why credit has or has not been awarded. In all cases, actual marks are indicated.

Questions worth 6 or more marks require you to:
- make connections between different parts of the subject content
- provide detailed explanations
- use examples and case studies to add geographical place detail
- back up your explanations with evidence.

It is worth thinking about the meaning of the command words you will encounter at A-level. The command words are given in the table below, by increasing level of demand. Higher-demand command words require higher-level thinking skills that include the ability to evaluate, draw conclusions and make judgements that are supported by evidence and make logical sense.

	Command word	Meaning	Marks
Increasing demand	Draw/plot/complete	Add information to correctly finish a graph, map, diagram or statistical test	1–4 marks
	Suggest	For an unfamiliar scenario, write a reasoned explanation of how or why something may occur. A suggested explanation requires justification/exemplification of a point that has been identified	3 or 4 marks
	Explain	Provide a reasoned explanation of how or why something occurs. An explanation requires understanding to be demonstrated through the justification or exemplification of points that have been identified	3, 4, 6 or 8 marks
	Assess	Use evidence to determine the relative significance of something. Give balanced consideration to all factors and identify which are the most important	12 marks
	Evaluate	Measure the value or success of something and ultimately provide a balanced and substantiated judgement/conclusion. Review information and then bring it together to form a conclusion, drawing on evidence such as strengths, weaknesses, alternatives and relevant data	20 marks

Tectonic processes and hazards

Question 1

(a) Study Figure 1. For each location shown in Figure 1, calculate the tsunami travel time from the origin to the location. (4 marks)

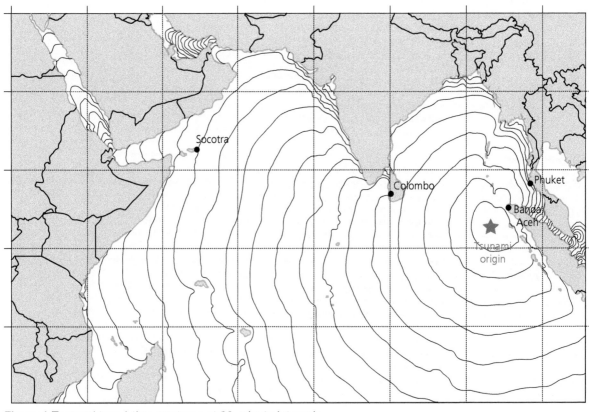

Figure 1 Tsunami travel-time contours at 30-minute intervals

(b) Assess the importance of prediction and forecasting in the successful management of tectonic disasters. (12 marks)

Part (a) is a skills question, using the information on Figure 1. The key here is to be accurate and to recognise that the isolines on Figure 1 are in 30-minute, not 1-hour, intervals. Don't rush these skills or 'doing' questions as mistakes are easy to make.

Part (b) is an extended writing question that is marked in levels. 'Assess the importance' means 'weighing up', so your answer needs to be evaluative in style. Both prediction and forecasting need to be mentioned, and high-quality answers will also consider other factors such as response and evacuation, as prediction and forecasting are not the only factors that determine whether tectonic disasters are successfully managed. Use examples to support your answer to give it the required depth. The levels mark scheme for 12-mark questions is shown below.

Level 1 1–4 marks	■ Demonstrates isolated elements of geographical knowledge and understanding, some of which may be inaccurate or irrelevant ■ Applies knowledge and understanding of geographical information/ideas, making limited logical connections/relationships ■ Applies knowledge and understanding of geographical information/ideas to produce an interpretation with limited relevance and/or support ■ Applies knowledge and understanding of geographical information/ideas to make unsupported or generic judgements about the significance of few factors, leading to an argument that is unbalanced or lacks coherence
Level 2 5–8 marks	■ Demonstrates geographical knowledge and understanding which are mostly relevant and may include some inaccuracies ■ Applies knowledge and understanding of geographical information/ideas logically, making some relevant connections/relationships ■ Applies knowledge and understanding of geographical information/ideas to produce a partial but coherent interpretation that is mostly relevant and supported by evidence ■ Applies knowledge and understanding of geographical information/ideas to make judgements about the significance of some factors, to produce an argument that may be unbalanced or partially coherent
Level 3 9–12 marks	■ Demonstrates accurate and relevant geographical knowledge and understanding throughout ■ Applies knowledge and understanding of geographical information/ideas logically, making relevant connections/relationships ■ Applies knowledge and understanding of geographical information/ideas to produce a full and coherent interpretation that is relevant and supported by evidence ■ Applies knowledge and understanding of geographical information/ideas to make supported judgements about the significance of factors throughout the response, leading to a balanced and coherent argument

Student answer

(a)

Location	Tsunami travel time	
	Hours	Minutes
Banda Aceh	1	30
Colombo	2	30
Phuket	2	30
Socotra	6	30

4/4 marks awarded. Part (a) scores 4 marks as all of the answers are accurate. There is some leeway in the minutes but not the hours.

(b) Prediction means the ability to say when and where a hazard will strike. For prediction to be useful it has to be accurate to within a few days, otherwise evacuations based on predictions become lengthy and problematic. Forecasting means giving the percentage chance of a hazard occurring ✓. For earthquakes, forecasting is possible ✓. For instance, the USGS forecasts that many California locations have a 10% chance of a 6.7 earthquake in the next 30 years ✓. This type of forecast quantifies risk, so is useful for people buying homes or for emergency services planning for a disaster. Earthquakes cannot be predicted, so prediction is of no use. Far more important is preparation and public education so people know what to do during and after an earthquake. Land-use zoning can be used to avoid building in areas of high risk, e.g. risk of liquefaction, and this can reduce economic and human losses. In contrast, volcanic eruptions can increasingly be predicted using gas spectrometers, tiltmeters and by analysing the seismic 'noise' made by mobile magma ✓. This allows for timely warning and evacuation. For instance 10,000s of people were saved by the timely evacuation around Mt Pinatubo when it erupted in 1991 ✓. However, monitoring volcanoes is costly and not done in all cases. This means that education, preparation and response should a disaster occur are important ✓. Technology such as ocean monitoring buoys can be used to predict the arrival of tsunami ✓, and even 5–10 minutes' warning can save large numbers of lives. However, it cannot prevent very high economic losses such as the US$150 billion in losses caused by the 2011 Sendai tsunami in Japan ✓. In conclusion, prediction is very important for eruptions and tsunami, but only forecasting is possible for earthquakes ✓. In all cases, prediction and forecasting tend to reduce human losses. Reducing economic loss depends more on long preparation and response.

12/12 marks awarded. The answer to part (b) is very good and scores 12 marks. It is well structured and uses good terminology. It shows a good understanding of both forecasting and prediction, recognising them as different things. There are examples used to support the answer, which provides depth. The answer is applied to volcanoes, earthquakes and tsunami so a good range of hazards is considered. The value of prediction is considered and judgements are made contrasting the case of volcanoes versus earthquakes. There is also assessment by considering that even though prediction is useful for eruptions, it is not available everywhere, meaning that other factors are important if disasters are to be successfully managed.

Question 2

(a) Study Table 1.

Table 1 Volcanic eruptions and death tolls in Indonesia

Death toll	Volcano name	Eruption date
5000	Kelud	1919
1584	Mount Agung	1963
1369	Mount Merapi	1930
426	Anak Krakatoa	2018
353	Mount Merapi	2010
18	Galunggung	1982
16	Sinabung	2014
7	Sinabung	2016
2	Kelud	2014
2	Mount Bromo	2004
Summary of data		
Number of eruptions	10	
Total death toll	8,777	

(i) Calculate the mean number of deaths recorded. (1 mark)

(ii) Calculate the median number of deaths recorded. (1 mark)

(iii) Calculate the interquartile range for the deaths recorded. You must show your working. (2 marks)

(b) Assess the value of earthquake hazard profiles in helping effective disaster management strategies. (12 marks)

> Part (a) is a skills question. It involves the use of a calculator, which you will need in the exam. You often have to do basic calculations but only for 1 or 2 marks. In (a)(iii) one of the marks is for showing your working, so you must not skip this stage. Only 1 mark is for the correct answer. Always give your answers to one decimal place.
>
> Part (b) is an extended answer in the form of an essay. Answers need to 'weigh up' different factors as part of a discussion. Stronger answers will make judgements throughout the answer in terms of which aspects of earthquake hazard profiles are most useful. 'Assess' questions are quite open: the answer needs to focus on the topic given (hazard profiles), but other factors (e.g. level of development) can be considered as part of a discussion. It is important to understand hazard profiles and the fact that they illustrate different aspects of an earthquake such as its duration, areal extent, predictability, frequency and speed of onset. You need to use examples to add depth to your answer. This question uses the same levels mark scheme as Question 1(b).

Student answer

(a) (i) 877.7

(ii) 18 + 353 = 371 divided by 2 = 185.5

(iii) 7 and 1369, so 1369 − 7 = 1362

(b) Hazard profiles show the key characteristics of earthquakes such as ✓ areal extent, magnitude, duration and predictability. Disaster management involves a range of strategies to reduce disaster impacts by managing before, during and after a hazard strikes. Hazard profiles represent the physical characteristics of a hazard but these are not the only factor that determines the success of management.

Magnitude ✓ is a key element of a hazard profile and in general larger magnitude earthquakes are more likely to overwhelm management strategies. Very large magnitude events like the 2011 Japanese ✓ earthquake and tsunami showed that even countries with very high levels of economic development cannot successfully manage a mega-disaster as deaths amounted to 18,000 and economic losses over $300 billion. Low-magnitude earthquakes can be managed by land-use planning ✓ and aseismic buildings but in low-income locations such as with the Nepal earthquake in 2015, poverty means that few people live in buildings that can resist ground-shaking and secondary hazards such as landslides.

> 4/4 marks awarded. The answers to part (a) are all correct so score 4/4. In (a)(ii) there is only a mark for the median value, not the calculation. However in (a)(iii) 1 mark for identifying the two quartiles and 1 mark for the final correct answer.

Immediate response and ✓ rescue are made easier if the ✓ areal extent of an earthquake is small, for instance the sequence of shallow earthquakes in ✓ Christchurch and Canterbury in New Zealand in 2010–11 was limited in areal extent, whereas the Kashmir and Sichuan earthquakes in 2005 and 2008 affected much larger areas. However, ✓ size or area affected is only one factor. The terrain in the Himalaya mountains hampered the relief efforts in 2005 and 2008 due to rugged mountains and isolated, inaccessible settlements.

If earthquakes are very infrequent, such as the Himalayan ones already referred to, there is a risk that community preparedness will be low because the risks are overlooked. In areas where there is a collective memory of a recent disaster, preparation and education are likely to be more thorough. It could be ✓ argued this was the case in Japan where the memory of previous earthquakes such as ✓ Kobe in 1995 and ✓ the Okushiri tsunami 1993 meant that preparation for the 2011 earthquake was reasonable: tsunami sirens ✓ were heeded by many and the death toll could have been much worse.

In conclusion ✓, frequency, magnitude and areal extent do have an effect on the success of hazard management. Magnitude is the most significant ✓ factor as very high-magnitude earthquakes can undo even the best laid plans. Level of development plays a crucial role as this human factor allows the investment in management strategies lacking in developing and some emerging countries. Physical factors such as terrain, not in the hazard profile, can also be ✓ important in determining how successful management and response are.

12/12 marks awarded.
The answer to part (b) fits the Level 3 descriptors because the answer shows good understanding of hazard profiles, including specific parts of the earthquake hazard profile. It applies examples to the concept of the hazard profile in order to show a more detailed understanding of it. The answer makes judgements about the importance of different aspects of the hazard profile in relation to management strategies. It uses evidence, in the form of examples and data, throughout the answer — this adds weight to the judgements and the conclusion. The answer finishes with a clear conclusion that ties the whole answer together, making it coherent.

Landscape systems, processes and change: Glaciated landscapes and change

Question 3

(a) Study Figure 2. Explain the evidence for recent glacier retreat in Figure 2. (6 marks)

Figure 2 The snout of the Höllentalferner glacier in Germany

This is a data stimulus question. The resource, in this case a photograph, needs to be studied carefully and evidence from it used to answer the question. The photograph shows a small glacier snout (lower, centre) with an extensive area of light-coloured sediment in front and at the side, with a wide and deep glacial valley in the background. These 6-mark questions in the Landscape systems, processes and change options are marked using the following levels mark scheme.

Level 1 1–2 marks	■ Demonstrates isolated or generic elements of geographical knowledge and understanding, some of which may be inaccurate or irrelevant ■ Applies knowledge and understanding to geographical information inconsistently. Connections/relationships between stimulus material and the question may be irrelevant
Level 2 3–4 marks	■ Demonstrates geographical knowledge and understanding which are mostly relevant and may include some inaccuracies ■ Applies knowledge and understanding to geographical information to find some relevant connections/relationship between stimulus material and the question
Level 3 5–6 marks	■ Demonstrates accurate and relevant geographical knowledge and understanding throughout ■ Applies knowledge and understanding to geographical information logically to find fully relevant connections/relationships between stimulus material and the question

Student answer

In Figure 2 a small glacier can be seen in the centre at the bottom. It is surrounded on either side and in front by light-coloured sediment. This is likely to be glacial moraine. It is recently formed because it is not covered in vegetation, unlike the valley floor in the distance. There are lines and ridges on the right in the moraine, which could be recent terminal moraine lines, suggesting the glacier was larger in the recent past. This glacial sediment has been dumped as the glacier has retreated towards the bottom of the photo. The steep rock valley sides on the left suggest the glacier was once much higher and filled the valley. Very bare, rough rock on the valley sides suggests recent glacial abrasion and plucking to create the craggy rock sides. The valley in the distance where the vegetation appears is roughly U-shaped, which could be explained by the ice filling the valley and extending over that area in the past. There is also an area that looks party vegetated but has no trees. This is probably the moraine from 50–100 years ago which has had time for some plants to grow in some areas.

6/6 marks awarded. The strength of this answer is its careful observation of the photograph. Features such as bare sediment, vegetated and partially vegetated areas are identified and related to possible past positions of the glacier. There is some interpretation and explanation of moraines, a key landform in identifying former ice positions, as well as an explanation of the shape of the valley and the possible past position of glacier ice. Good understanding of glacial erosion processes is shown in terms of their impact on the valley shape and features.

(b) Explain the importance of freeze–thaw weathering in the formation of periglacial landforms.

(6 marks)

Part (b) is a level-marked question (see levels mark scheme on p. 155). Answers need to demonstrate an understanding of frost shattering and link it to landforms (plural). Questions such as this, which focus on physical processes, need good use of terminology and logical, sequential explanations.

Student answer

Freeze–thaw weathering is one of the most important physical processes in periglacial areas. It relies on repeated freezing–thawing cycles, which is a feature of the periglacial climate. Water freezes in cracks and expands by 9% in volume, exerting a force that can split rocks into angular fragments. These angular fragments are the basis of patterned ground formation as the related process of frost heave moves larger freeze–thaw weathered rocks towards the surface to form striped polygons. Frost heave is the repeated expansion and contraction of the permafrost layer from season to season. Most periglacial landforms, such as solifluction lobes and terraces, and blockfields, consist of freeze–thaw debris.

6/6 marks awarded. The answer uses good terminology throughout and explains the process of freeze–thaw weathering sequentially. There is detailed understanding, such as reference to water expanding by 9% on freezing. The process is then linked to the related process of frost heave and then to a number of named landforms, i.e. patterned ground, solifluction lobes and blockfields. It is a good answer because it recognises that freeze–thaw is one of a number of processes operating in tandem in periglacial areas rather than just considering it in isolation.

(c) Explain how glacial landforms can be used to help reconstruct former ice mass movement and extent.

(8 marks)

This style of question is best thought of as a 'mini essay'. The topic is quite a narrow one, focused on a small area of physical geography. There is no stimulus material so you need to use detailed knowledge and understanding to provide explanations. Examples, but not large case studies, can also be used to support your explanations. In this question answers need to explain both ice mass extent and movement. These 8-mark questions in the Landscape systems, processes and change options are marked using the following levels mark scheme.

Level 1 1–2 marks	■ Demonstrates isolated elements of geographical knowledge and understanding, some of which may be inaccurate or irrelevant ■ Understanding addresses a narrow range of geographical ideas, which lack detail
Level 2 3–5 marks	■ Demonstrates geographical knowledge and understanding which are mostly relevant and may include some inaccuracies ■ Understanding addresses a range of geographical ideas, which are not fully detailed and/or developed
Level 3 6–8 marks	■ Demonstrates accurate and relevant geographical knowledge and understanding throughout ■ Understanding addresses a broad range of geographical ideas, which are detailed and fully developed

Student answer

Ice masses include glaciers and ice sheets. The former extent of large ice sheets can be partly indicated by isostatic readjustment ✓ of the land surface which was depressed by the weight of ice. This method is only useful for large ice sheets, not valley glaciers ✓. Far more accurate is the use of terminal moraines ✓. These mounds of glacial sediment can indicate the maximum extent of glacier and ice sheets. The Valparaiso moraine south of Chicago indicates the maximum southward extent of the last glacial period. Other landforms, such as eskers ✓, are known to have formed englacially or subglacially ✓ so can indicate areas that had ice cover. Eskers may also indicate flow direction assuming meltwater and ice flowed in similar directions. Large ice sheets have a meltwater plain or sandur in front of the ice ✓ which consists of meltwater channels and water-deposited sediment. The edge of this sandur can be used to infer the edge of the ice. One of the most useful indicators of ice flow direction are erratics ✓. These are rocks and boulders deposited in a location distant from their source rock because the rocks were carried by ice. If the source outcrop can be identified then the flow direction of ice can be inferred. Striations ✓, which are grooves cut into rock by abrasion ✓, also indicated ice flow direction but only over short distances ✓.

8/8 marks awarded. This is a Level 3 answer. It uses good terminology about glacial processes. A range of different landforms is considered and their role is explained. The answer relates to both ice extent and ice movement and has more than one example of a landform for each, so the answer has a good range of ideas. There are some comments on the usefulness of the landforms, which, while not directly asked for, help demonstrate depth of understanding.

(d) Evaluate the extent to which management can balance the demands of conservation and economic development in glacial environments. (20 marks)

This question is an essay question with a high mark tariff. The question has a number of different elements to it (conservation + economic development), all of which need to be covered. There are concepts which need to be addressed too (management, demands). The command word 'evaluate' means 'weigh up and come to a judgement'. Good answers will argue a case using examples and case studies to back up the argument. For this question, it might be argued that the balance is possible to achieve, but that is not the case everywhere, i.e. in areas with high pressure versus more isolated areas with few inhabitants or visitors. These 20-mark questions in the Landscape systems, processes and change options are marked using the following levels mark scheme.

Level 1 **1–5 marks**	■ Demonstrates isolated elements of geographical knowledge and understanding, some of which may be inaccurate or irrelevant ■ Applies knowledge and understanding of geographical ideas, making limited and rarely logical connections/relationships ■ Applies knowledge and understanding of geographical information/ideas to produce an interpretation with limited coherence and support from evidence ■ Applies knowledge and understanding of geographical information/ideas to produce an unsupported or generic conclusion, drawn from an argument that is unbalanced or lacks coherence
Level 2 **6–10 marks**	■ Demonstrates geographical knowledge and understanding which are occasionally relevant and may include some inaccuracies ■ Applies knowledge and understanding of geographical information/ideas with limited but logical connections/relationships ■ Applies knowledge and understanding of geographical ideas in order to produce a partial interpretation that is supported by some evidence but has limited coherence ■ Applies knowledge and understanding of geographical information/ideas to come to a conclusion, partially supported by an unbalanced argument with limited coherence
Level 3 **11–15 marks**	■ Demonstrates geographical knowledge and understanding which are mostly relevant and accurate ■ Applies knowledge and understanding of geographical information/ideas to find some logical and relevant connections/relationships ■ Applies knowledge and understanding of geographical ideas in order to produce a partial but coherent interpretation that is supported by some evidence ■ Applies knowledge and understanding of geographical information/ideas to come to a conclusion, largely supported by an argument that may be unbalanced or partially coherent
Level 4 **16–20 marks**	■ Demonstrates accurate and relevant geographical knowledge and understanding throughout ■ Applies knowledge and understanding of geographical information/ideas to find fully logical and relevant connections/relationships ■ Applies knowledge and understanding of geographical information/ideas to produce a full and coherent interpretation that is supported by evidence ■ Applies knowledge and understanding of geographical information/ideas to come to a rational, substantiated conclusion, fully supported by a balanced argument that is drawn together coherently

Student answer

There are some examples of glacial environments where conservation of biodiversity and the unique landscape are placed well ahead of desires to develop areas for their economic resources. The most famous example is Antarctica ✓ which has been protected under the Antarctic Treaty since 1961 ✓. All commercial exploitation of mineral resources is banned and fishing and tourism are strictly regulated. This works because there is an international consensus that Antarctica ✓ is a unique, pristine environment that should be the preserve of science, not economic development. Elsewhere views are not so clear cut ✓. The periglacial tundra of northern Alaska contains the Arctic National Wildlife Reserve (ANWR) ✓. This is a unique landscape of glacial mountains and permafrost plains with a large caribou and polar bear population. It is also thought to contain rich oil and gas reserves. Currently it is a strictly protected wilderness ✓ but this may change as pressure from oil companies and some Alaskan residents grows to tap the economic potential of the area ✓.

The Antarctic and ANWR are remote, isolated active glacial environments. Relict ✓ glacial environments such as the UK's Lake District ✓ are much more accessible and consequently are under pressure from tourism. Tourism contributes up to 40% of the area's economy and developing it creates jobs and economic development. However, fragile uplands are prone to trampling, litter and footpath erosion, and the area's carrying capacity may even be exceeded in summer months ✓. The Lake District National Park authority manages the area using the 'Sandford Principle', meaning that conservation takes priority over economic development if the two conflict ✓. While this helps preserve the glacial landscape, it risks making people dependent on a small number of low-paid, seasonal jobs ✓. It also means large developments such as HEP reservoirs and quarries have no chance of being developed. In the Alps ✓, which has active and relict ✓ landscapes, the fact that managing the area involves numerous different countries has meant that damage to the landscape from ski resorts, motorways and urbanisation has occurred despite the Alpine Convention that should protect the area. Overall, a balance is easier to strike in extreme, isolated environments with few or no residents. It is much harder in accessible locations and where more than one country is involved ✓.

20/20 marks awarded. This answer is Level 4. It uses a wide range of examples, which are applied to the concepts of economic development and conservation. The range of examples includes both active and relict landscapes, giving good balance. The examples include details of how the areas are managed There are also judgements made about the success of the different management strategies. The overall approach is evaluative, with different views being presented to produce a coherent overview. There is a clear overall judgement that draws together the evidence presented.

Landscape systems, processes and change: Coastal landscapes and change

Question 4

(a) Study Figure 3. Explain the formation of the landforms shown.   (6 marks)

Figure 3 Landforms on the Costa Brava, Spain

> This is a data stimulus question. The resource, in this case a photograph, needs to be studied carefully and evidence from it used to answer the question. The photograph shows a small rock arch in a cliff. There is also a cave feature beneath and to the right of the arch. In several places, e.g. around the cave mouth, a small wave-cut notch can be seen. A small rock stump can be seen in the foreground.

Student answer

The main landform in Figure 3 is a rock arch. This formed by abrasion and hydraulic action. The rock in Figure 3 looks like heavily fractured limestone. Hydraulic action would happen when incoming waves crashed into the cliff face, pressurising air in the fractures and forcing them apart. Combined with the sand-papering action of abrasion as waves hurl sediment at the rock, the erosion has forced through a small headland, forming an arch. The top of the arch is protected from direct wave erosion due to its height. To the right of the arch is a cave. This is the first stage of arch formation. The cave could form where a weakness like a small fault or large fracture creates an area of weak rock that abrasion and hydraulic action can exploit. The high temperatures in Spain may mean some chemical weathering has also contributed to the landforms, e.g. by widening joints and fractures.

6/6 marks awarded. This answer uses good terminology to explain the erosion processes that formed the arch, and makes good use of Figure 3. Observations from the photograph are accurate and the answer picks out and explains details of the rock type and smaller landforms such as the cave. There is detailed understanding of the impact of weaknesses in the rock, and how abrasion and hydraulic action contribute to the formation of both the arch and the cave. Some wider understanding is shown by the suggestion that weathering would also be a contributing factor.

(b) Explain why it is difficult to predict future sea levels. (6 marks)

Part (b) is a levels-marked question (see levels mark scheme on p. 155). A range of explanations is needed that might explain why predicting sea levels in the future is difficult. These could be either physical processes or human issues.

Student answer

Sea levels are hard to predict for two main reasons. Firstly, it is hard to know how ice stores will respond to global warming. Small differences in future global temperatures could result in large changes to melting rates on Greenland and Antarctica. Most sea level rise between 1990–2010 was due to thermal expansion increasing the volume of ocean water. Future thermal expansion depends on future temperature, which can't be known. In some locations such as Norway and Scotland, isostatic change is happening alongside eustatic sea level rise, making prediction complex. Secondly, because future sea levels depend on future temperature, predictions are only possible if future temperatures are known. They are not because they depend on future greenhouse gas levels and the amount of these depends on other unknowns such as future global population, affluence, fossil fuel use as well as attempts to reduce emissions.

6/6 marks awarded. The answer focuses on the uncertainty of future sea levels in terms of physical processes such as ice melt and human activities that could affect the rate of future planetary warming. The range of explanations is good, as is the use of terminology which is precise and shows understanding (eustatic, isostatic, thermal expansion). There is understanding of the links between human actions and physical systems and some located examples are used to support the explanation.

(c) Explain how bedrock lithology and geological structure can influence rates of coastal recession. (8 marks)

This style of question is best thought of as a 'mini essay'. The topic is quite a narrow one, focused on a small area of physical geography. There is no stimulus material so you need to use detailed knowledge and understanding to provide explanations. Examples, but not large case studies, can also be used to support your explanations.

Student answer

Lithology has a significant impact on rates of coastal recession. Resistant rocks often form headlands. An example is Flamborough Head in Yorkshire ✓. The chalk here erodes at 1–2 mm per year through abrasion and hydraulic action ✓. South of Bridlington the boulder clay of the Holderness coast erodes at 2–3 m per year because it is much less resistant to erosion ✓. Boulder clay also recedes due to mass movements, especially rotational slides. These are often linked to storm events which undermine the cliffs but also saturate them with heavy rain, leading to internal failure. Lulworth Cove shows how rock type and structure can influence ✓. The cove is part of the concordant Dorset coast ✓. Resistant beds of Portland and Purbeck limestone form the narrow cove entrance, and hard chalk the steep cliff at the back of the cove ✓. The wide part of the

8/8 marks awarded. This is a Level 3 answer. The answer uses a number of real examples that add depth to the explanation. The importance of lithology (rock type) is explained with reference to these examples and some data are provided on recession rates. The Lulworth Cove example combines rock type and geological structure, showing that the two cannot be dealt with separately. Some further details of minor structural features are also explained. The answer has a good range of ideas for both lithology and structure.

cove has been eroded from softer clays, which are less resistant to erosion. On the Northumberland coast, rates of recession are often determined by the weakest strata in a cliff ✓. Often this is a coal seam, which readily forms a wave-cut notch, leading to the collapse of more resistant rocks above ✓. Faults and large joints are often preferentially eroded as they are weaker than surrounding rocks and become the locations for caves and arches, eroding more quickly than surrounding rocks ✓.

(d) Evaluate the extent to which the twin threats of sea level rise and erosion on coastlines can be managed in a sustainable way. (20 marks)

This is an essay question with a high mark tariff. The question has a number of different elements to it (sea level rise + erosion + management), all of which need to be covered. There are concepts that must be addressed too (threat, sustainable). The command word 'evaluate' means 'weigh up and come to a judgement'. Good answers will argue a case using examples and case studies to back up the argument. For this question it might be argued that erosion can be managed in a sustainable way as it is a local threat, whereas sea level rise is much harder to manage as sea level is rising everywhere (although it threatens some coasts much more than others).

Student answer

Sustainable management of coasts means managing littoral cells holistically rather than managing isolated pockets of coast ✓. It also means management that meets the social and economic needs of residents and users as well as maximising environmental protection. Erosion threats can be managed using Integrated Coastal Zone Management (ICZM) and Shoreline Management Plans (SMPs) ✓. In the UK SMPs are drawn up by local authorities working collaboratively, and with all stakeholders, to manage a long stretch of coastline. Decisions about which areas to protect from erosion, i.e. hold the line, are made based on the cost and technical feasibility of defences versus the value of what is being protected. On the Holderness coast towns like Hornsea and Withernsea are protected but low-value farmland is not ✓. The overall SMP works with the natural Holderness sediment cell to maximise protection in some places, with the overall aim of reducing erosion rates over a timescale of 100 years or more ✓.

Erosion management is fairly local in scale and the threatened areas tend to be small ✓. Sea level rise is a widespread threat and as such is much harder to manage. It threatens London, as well as the Essex Marshes, large areas of Norfolk and Lincolnshire and many other areas ✓. It cannot be managed at a local level only because the basic cause of most sea level rise is global warming. This problem requires action at a global scale to reduce emissions. Local areas are left to deal with the consequences of sea level rise but cannot manage

20/20 marks awarded. This is a Level 4 answer. It shows good conceptual understanding by defining sustainable management at the start and relating this to specific strategies. A range of real examples is used with some detail, which provides the evidence needed to make a judgement. The Holderness example is used to show that if management is holistic, it can be sustainable.

There is a good balance between the sections on erosion and sea level rise. There is an evaluative comparison made between the two threats, contrasting the extent to which they can be managed.

the fundamental cause. However, the threat can be managed in a sustainable way to some extent. The case of Abbotts Hall Farm on the Blackwater Estuary in Essex shows that numerous stakeholders can agree even when difficult decisions need to be made ✓. In 2002 five breaches in the sea wall here turned 4000 hectares of land into a managed realignment scheme. This scheme removed the risk of sea level rise by turning a once-protected area into new salt marshes, showing that environmentalists, landowners, coastal managers and local people and businesses can all be kept happy even when radical plans are adopted. Overall, managing coastal erosion in a sustainable way is possible if it is done in a holistic way. There will always be winners and losers but decisions can ensure most people are winners. Sea level rise is more about managing consequences than causes and as such is harder, but even radical new ideas can lead to sustainable outcomes ✓.

> The Blackwater Estuary case study is used very well to show that even the more difficult problem of sea level rise can be managed sustainably. The overall conclusion is supported by the evidence presented.

Physical systems and sustainability

Question 5

(a) Study Figure 4. Suggest **one** reason for the trend in global methane atmospheric concentration.

(3 marks)

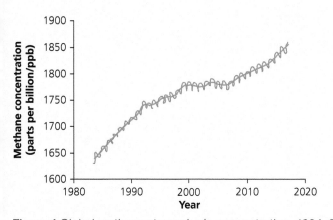

Figure 4 Global methane atmospheric concentration, 1984–2020

> This is a data stimulus question. The first task is to recognise the upward trend on Figure 4 and the fact that a steep increase paused around 2000 but resumed after 2008. The question demands **one** reason. This means identifying a reason for increased methane concentrations and explaining it with two further extended points. You cannot name multiple reasons, such as an increase in cattle farming and an increase in methane emission from landfill. Stick with the first reason and give additional detail.

Student answer

The trend in methane levels is a steep rise from 1625 ppb in 1984 to 1850 ppb by 2015, including a slower growth phase from 2000–2008. The main cause of this is agriculture and especially the expansion of cattle farming. It is released by the growing number of cattle needed to meet demand for beef in emerging countries. Expanded cattle ranches often replace carbon-sequestering forests with methane-emitting cattle.

3/3 marks awarded. The first sentence does not score marks because it is a description; however, it does provide a focus for the explanation that follows. The one reason given — agriculture — is expanded on in terms of cattle farming, beef demand and replacing forest with ranches. The last sentence shows a good understanding of the carbon cycle.

(b) Explain the role of the oceans in the carbon cycle.

(6 marks)

This is a levels-marked question that requires an in-depth explanation of one part of the carbon cycle. It might be tempting to discuss other parts of the cycle, but this will not gain marks unless there is a link to the oceans. Focus on the oceans and link the ocean to other relevant parts of the cycle such as the atmosphere and geological stores. This question is quite a conceptual one that demands good use of carbon cycle terminology, i.e. you need to have revised the physical processes in depth. It is much less about 'place' because it is focused on a cycle. These 6-mark 'explain' questions in A-level Paper 1 are marked using the following levels mark scheme.

Level 1	1–2 marks	Demonstrates isolated elements of geographical knowledge and understanding, some of which may be inaccurate or irrelevant
		Understanding addresses a narrow range of geographical ideas, which lack detail
Level 2	3–4 marks	Demonstrates geographical knowledge and understanding which are mostly relevant and may include some inaccuracies
		Understanding addresses a range of geographical ideas, which are not fully detailed and/or developed
Level 3	5–6 marks	Demonstrates accurate and relevant geographical knowledge and understanding throughout
		Understanding addresses a broad range of geographical ideas, which are detailed and fully developed

Student answer

The oceans contain about 38,000 gigatonnes of carbon, making them Earth's largest carbon store. Carbon is stored as dissolved CO_2 in water, or stored in algae, plants and coral. The biological pump sequesters carbon from the atmosphere through the growth of phytoplankton (photosynthesis). These form the basis of the ocean food web. Passed along the food web, carbon is returned to the atmosphere by biological decay. The biological pump takes place on a timescale of hours to years and the flux between atmosphere and ocean is 11 gigatonnes per year. Only a small proportion of carbon enters the carbon pump. This is when dead organic material, e.g. shells and dead phytoplankton, sinks to the ocean floor and becomes

6/6 marks awarded. The strength of this answer is its accurate use of the correct terminology of the carbon cycle. The simplest part of the oceanic carbon cycle is phytoplankton taking in atmospheric carbon during photosynthesis. The answer goes beyond that to explain the role of the carbonate and physical pumps, not just the biological pump. The answer shows understanding of carbon stores and fluxes, as well as the fact that these operate on different timescales. Carbon in the oceans is linked to atmospheric and geological stores, but the focus of the answer remains on oceanic carbon. Notice that some data values are used. These are very useful in showing depth of understanding.

sediment that will eventually become carbonate rocks like limestone (geological carbon store), although on timescales of millions of years. Within the oceans, the physical pump, in the form of ocean currents and the thermohaline circulation, moves carbon vertically and horizontally. The size of the ocean carbon store makes it very important, as it has the capacity to store excess carbon produced by human activities.

Question 6

(a) Study Figure 5. Explain **one** reason why some European countries will have high levels of water stress by 2040.

(3 marks)

Water stress in 2040
- Low
- Low to medium
- Medium to high
- High
- Extremely high

Figure 5 Water stress levels in Europe in 2040

This question refers to Figure 5, the map of water stress in Europe by 2040. The risk with this question is that you explain why two or three different places might have extremely high water stress by 2040, e.g. Spain, Greece and Italy. What is actually required is one in-depth explanation using extended points. There are several to choose from, including reduced supply because of global warming-induced climate change, or increased demand from rising populations of specific economic activities such as tourism.

Student answer

Mediterranean countries such as Portugal, Spain, Italy and Greece will experience high water stress by 2040 because of climate stress due to global warming. Higher global temperatures will lead to greater evaporation and smaller soil and surface water stores. In addition, shifts in climate belts will mean lower and more unreliable rainfall as climates become more arid, reducing water availability and increasing water stress.

3/3 marks awarded. This is a good answer that takes the idea of global warming as the cause and expands on it in terms of greater evaporation, changes to rainfall patterns and reduced water stores. It is written as a logical sequence of points. Because it uses terminology from the water cycle, it shows understanding of that concept. The key here is to write an extended explanation of a single cause — which this answer does.

(b) Explain how human actions can increase flood risk. (6 marks)

This question is focused on human actions, not physical processes. The latter are relevant, but only when linked to a change brought about by human activity — such as decreased infiltration resulting from deforestation. This question is slightly different when compared with part (a) because it would benefit from place-specific detail, i.e. named places where human activity has increased flood risk. Ideally, you should explain two or three human actions in detail rather than try to cover four or five. That approach will tend to produce a descriptive list rather than an explanation (see the mark scheme levels grid for Question 1(b)).

Student answer

Human activity alters the hydrological cycle in a number of ways that can increase flood risk. Widespread urbanisation contributed to flooding in the UK in summer 2007. Summer rainfall totals were the highest for 200 years, but impermeable urban surfaces and urban drainage increased surface runoff in cities such as Sheffield and Hull, causing river hydrographs to have reduced lag times and higher peaks. Widespread construction on floodplains decreases their flood storage capacity and places properties at higher risk from even minor flooding. Deforestation reduces interception which slows precipitation's journey to the surface. This in turn reduces infiltration and increases surface runoff. Removal of trees also reduces evapotranspiration, meaning more water enters river systems by surface runoff. Steep slopes in Manila, Philippines, are deforested to create new urban slums. These are prone to flooding and landslides, especially on steep slopes which promote surface runoff. Some estimates suggest that urbanising a previously forested area increases surface runoff from 10% to 55% of precipitation and reduces evapotranspiration from 40% to 30%.

6/6 marks awarded. This answer focuses in detail on two explanations: urbanisation and deforestation. This is a good approach, especially because some place detail is added for both, which shows the student understands flood risk in contrasting locations. The answer recognises that the drainage basin hydrological cycle is the key to understanding increased flood risk. The terminology used focuses on the pathways water takes between reaching the ground and reaching a river channel. There is also a useful link between deforestation and urbanisation in the context of Manila. This answer manages to avoid being a list of problems and instead is a detailed explanation.

Question 7

Explain why there is uncertainty about future greenhouse gas levels in the atmosphere. (8 marks)

This question needs a detailed understanding of the concept of 'uncertainty' in terms of both human and physical processes. Aim to explain a range of reasons rather than only one or two. This could include uncertainty about future population and affluence levels, as well as uncertainty resulting from mitigation attempts. Physical feedback mechanisms are also relevant. These 8-mark 'explain' questions in A-level Paper 1 are marked using the following levels mark scheme.

Level 1	1–3 marks	Demonstrates isolated elements of geographical knowledge and understanding, some of which may be inaccurate or irrelevant
		Understanding addresses a narrow range of geographical ideas, which lack detail
Level 2	4–6 marks	Demonstrates geographical knowledge and understanding which are mostly relevant and may include some inaccuracies
		Understanding addresses a range of geographical ideas, which are not fully detailed and/or developed
Level 3	7–8 marks	Demonstrates accurate and relevant geographical knowledge and understanding throughout
		Understanding addresses a broad range of geographical ideas, which are detailed and fully developed

Student answer

Carbon dioxide levels in 2017 were 400 ppm ✓, an increase from 315 ppm in the late 1950s. IPCC projections for 2100 range from 500 ppm to 900 ppm, showing there is a large amount of uncertainty ✓.

The first reason ✓ for uncertainty is because greenhouse gas emissions of carbon dioxide, methane and nitrogen oxides are caused by human activities such as farming, industry, electricity generation (fossil fuels) and transport. These emissions result from resource consumption. This is related to both total population and the average level of affluence ✓. UN population projections for 2100 range from 8 to 12 billion. Not only is the future number of people not known, but their wealth and ecological footprint can't be known either.

Secondly, efforts to mitigate greenhouse gas emissions could reduce the growth rate of emissions. The 2015 COP21 meeting in Paris ✓ agreed some emissions targets but these may not be met. On the other hand, future agreements could reduce emissions further ✓. A business-as-usual emissions scenario would probably mean carbon dioxide levels of 800–900 ppm by 2100, i.e. people consume resources, especially fossil fuels, in much the same way in the future as they do today.

Thirdly, the response of physical systems such as the carbon cycle to global warming is uncertain. Global warming could increase melting of Arctic permafrost ✓, which would lead to biological decay of frozen organic material releasing carbon dioxide. In addition, it would release trapped methane — a powerful greenhouse gas. On the other hand, warmer temperatures could lead to increased forest growth, which would sequester carbon from the atmosphere ✓. The ocean is the

8/8 marks awarded. These 8-mark questions sit between the 6-mark 'explain' and 12-mark 'assess' questions. They benefit from being organised carefully, and you might want to consider an introductory sentence to focus your answer. In this case there is a starting focus on carbon dioxide levels with some useful specific data. The answer then clearly states the future uncertainly over carbon dioxide levels. The structure for the answer comes from three paragraphs, each focused on a different explanation. This gives the answer good range. The first paragraph focuses on emissions uncertainty, recognising that this is complicated by not knowing either total future population or resources consumption per person. There is some depth on climate mitigation agreements as well as recognition that future agreements cannot be known, therefore creating more uncertainty. The third paragraph broadens the explanation to include physical systems uncertainty, and recognises physical systems could potentially release or store more greenhouse gas emissions.

Earth's largest carbon store that has a relatively rapid rate of flux. The oceans could sequester more carbon from the atmosphere in the future but this is uncertain ✓.

In summary ✓, the number and wealth of humans, how humans choose to use resources and which resources they use, as well as the response of physical systems to a warmer world, are all uncertain.

> The summary is a good way to round off the answer. A judgement is not really needed because the command is 'explain', but the summary completes a well-structured answer.

Question 8

Study Table 2. Assess the costs and benefits of Singapore's planned water supply changes by 2060.

(12 marks)

Table 2 Key facts about Singapore's population, water use and water supply in 2016 and planned by 2060

Year	Total population	Water use per person per day (litres)	Four sources of Singapore's fresh water			
			Recycled, treated, grey water ('NEWater')	Desalination	Transboundary water imports from Malaysia	Local rainfall catchments
2016	5.69 million	150	30%	10%	40%	20%
2060 planned	6.56 million	130	55%	30%	–	15%

> This question is a data stimulus question. Your answer must refer in detail to the information in Table 2: quote it in your answer. Table 2 needs some careful analysis first. It shows a major change in how Singapore will get its water by 2060, with some sources increasing and others even disappearing. Both costs and benefits need to be covered in a balanced way. Often recalling the words 'social, economic, environmental and political' helps to structure an answer to a costs and benefits question (or advantages and disadvantages). As the command is 'assess', your answer needs to weigh up the different costs and benefits and must start to consider whether some are more significant than others. This is required for a good Level 3 answer. These 12-mark 'assess' questions in A-level Paper 1 are marked using the following levels mark scheme.

Level 1	1–4 marks	■ Demonstrates isolated elements of geographical knowledge and understanding, some of which may be inaccurate or irrelevant ■ Applies knowledge and understanding of geographical information/ideas, making limited logical connections/relationships ■ Applies knowledge and understanding of geographical information/ideas to produce an interpretation with limited relevance and/or support ■ Applies knowledge and understanding of geographical information/ideas to make unsupported or generic judgements about the significance of few factors, leading to an argument that is unbalanced or lacks coherence

Level 2	5–8 marks	▪ Demonstrates geographical knowledge and understanding which are mostly relevant and may include some inaccuracies
		▪ Applies knowledge and understanding of geographical information/ideas logically, making some relevant connections/relationships
		▪ Applies knowledge and understanding of geographical information/ideas to produce a partial but coherent interpretation that is mostly relevant and supported by evidence
		▪ Applies knowledge and understanding of geographical information/ideas to make judgements about the significance of some factors, to produce an argument that may be unbalanced or partially coherent
Level 3	9–12 marks	▪ Demonstrates accurate and relevant geographical knowledge and understanding throughout
		▪ Applies knowledge and understanding of geographical information/ideas logically, making relevant connections/relationships
		▪ Applies knowledge and understanding of geographical information/ideas to produce a full and coherent interpretation that is relevant and supported by evidence
		▪ Applies knowledge and understanding of geographical information/ideas to make supported judgements about the significance of factors throughout the response, leading to a balanced and coherent argument

Student answer

Table 2 shows that Singapore plans to dramatically shift its water supply from imports from Malaysia towards water recycling and desalination by 2060 ✓. At the same time its population will rise by close to 1 million but water use per person falls by 20 litres per day.

There is potentially a very large benefit of ending use of imported water from Malaysia ✓. In 2016 40% of Singapore's water came from another country that could reduce or stop the supply. There are examples where water-sharing agreements lead to tensions, such as in the Nile basin ✓, so relying on another country for water is potentially risky. However, if the relationship is just an economic one, i.e. Singapore pays Malaysia for water, it could be sustainable ✓. It depends on the price Singapore pays per litre, which could be high and is controlled by Malaysia.

Recycled grey water is an example of water conservation. It has the benefits of using the same water multiple times, so reducing demand for new supply. Costs might include complex and expensive collection and treatment systems, plus the public in Singapore may not like the idea of 'NEWater' that has been recently used in

12/12 marks awarded. This is a good answer. It refers to all of the data in Table 2. The first paragraph makes direct reference to the data and this shows good understanding of it. At the start, there is a clear judgement about the significance of the benefit of Singapore reducing its reliance on water from Malaysia. Some brief reference is made to other transboundary water situations, which shows breadth of understanding. Assessment is shown by the counter-argument that getting water from Malaysia may not actually be very insecure at all. The paragraph on grey water recycling/NEWater also covers costs and benefits, showing that the issue is being considered from both sides.

someone's shower. Water conservation is also evident in the aim to reduce per capita daily consumption. In the UK this is achieved by more efficient washing machines, dishwashers and showers. It may have the benefit of reducing water bills, as well as preventing increases in water demand ✓.

The shift to 30% of water from desalination has the benefit that Singapore will control this water supply, i.e. rather than Malaysia controlling it. However, desalination has costs ✓. It is expensive to build the plants, and these need a large energy source such as oil or natural gas to run. This means desalination is usually not eco-friendly and has high greenhouse gas emissions. It may make water bills higher because of the added production costs. However, it does ensure water security.

Rainfall catchments are set to decline by 5% ✓. This might actually mean the volume of water they supply is similar to 2016 because despite the 150–130 litre reduction in per person water use, total water demand will rise because of the 800,000 increase in population.

Overall, by 2060 Singapore will have the very significant benefit of being water secure and no longer relying on Malaysia. However, this could come at the economic costs of higher water bills and some environmental costs from desalination ✓.

> Environmental and economic costs are considered in relation to desalination, as well as benefits in terms of a secure water supply. The section on rainfall catchments is very analytical, i.e. it unpicks the data to understand the changing picture of Singapore's water demand and shows good understanding. When the command is 'assess', a conclusion is always useful, in this case stressing the primary benefit of greater water security for Singapore despite some potential costs.

Question 9

Evaluate the view that renewable energy sources can easily meet future global energy demand.

(20 marks)

> This is an essay question. As such, detailed place knowledge and understanding in the form of examples and case studies are important. A very good answer cannot be written in only general terms. Answers need to show understanding of what future energy demand might be, as well as a range of renewable sources (wind, solar, HEP) rather than only one in detail. The key to a high mark is understanding that 'evaluate' requires supported judgements to be made. For example, could wind power meet a significant chunk of future energy demand or is it too physically constrained, expensive and intermittent to do so reliably in many places? You need to ask yourself these questions and provide the answers. Notice that the word 'easily' appears in the question and strong answers will consider how far easily is true, for different energy sources. Good answers will make conclusions based on different energy sources and different locations. These 20-mark 'evaluate' questions in A-level Paper 1 are marked using the following levels mark scheme.

Level 1	1–5 marks	▪ Demonstrates isolated elements of geographical knowledge and understanding, some of which may be inaccurate or irrelevant ▪ Applies knowledge and understanding of geographical ideas, making limited and rarely logical connections/relationships ▪ Applies knowledge and understanding of geographical information/ideas to produce an interpretation with limited coherence and support from evidence ▪ Applies knowledge and understanding of geographical information/ideas to produce an unsupported or generic conclusion, drawn from an argument that is unbalanced or lacks coherence
Level 2	6–10 marks	▪ Demonstrates geographical knowledge and understanding which are occasionally relevant and may include some inaccuracies ▪ Applies knowledge and understanding of geographical information/ideas with limited but logical connections/relationships ▪ Applies knowledge and understanding of geographical ideas in order to produce a partial interpretation that is supported by some evidence but has limited coherence ▪ Applies knowledge and understanding of geographical information/ideas to come to a conclusion, partially supported by an unbalanced argument with limited coherence
Level 3	11–15 marks	▪ Demonstrates geographical knowledge and understanding which are mostly relevant and accurate ▪ Applies knowledge and understanding of geographical information/ideas to find some logical and relevant connections/relationships ▪ Applies knowledge and understanding of geographical ideas in order to produce a partial but coherent interpretation that is supported by some evidence ▪ Applies knowledge and understanding of geographical information/ideas to come to a conclusion, largely supported by an argument that may be unbalanced or partially coherent
Level 4	16–20 marks	▪ Demonstrates accurate and relevant geographical knowledge and understanding throughout ▪ Applies knowledge and understanding of geographical information/ideas to find fully logical and relevant connections/relationships ▪ Applies knowledge and understanding of geographical information/ideas to produce a full and coherent interpretation that is supported by evidence ▪ Applies knowledge and understanding of geographical information/ideas to come to a rational, substantiated conclusion, fully supported by a balanced argument that is drawn together coherently

Student answer

Global energy demand is expected to increase by 40–50% from today until 2040 ✓. Almost all of this increased demand is driven by emerging and developing countries. Projections from the US Energy Information Administration suggest there will be almost no growth in energy consumption in developed countries. This means there are really two questions: ✓ can renewable energy easily replace fossil fuels used in the developed world, and can renewable energy be the major source of future energy supply in other countries?

Globally, 80% of energy production today is from fossil fuels. In developed countries this is changing in two ways. Firstly, a switch from coal to natural gas, which is cleaner. Secondly, a shift to renewables, especially wind power. Denmark gets 40% of its electricity from wind power and Germany 10%, but globally it only accounts for 4% of electricity ✓. Wind's intermittent and unreliable nature means that standby power stations (usually gas-fired) need to be available to boost supply, which adds to costs and makes the switch less easy than might be thought. Hydroelectric power (HEP) has been widely developed in some countries ✓. However, it requires specific geographical conditions, i.e. a reliable water supply and valleys that can be flooded to create HEP reservoirs. Canada, Brazil and Norway all generate over 50% of their electricity from HEP. Most developed countries have already utilised suitable sites and have limited capacity to expand so this may not be easy at all. This is not the case in some developing and emerging countries such as Ethiopia and China, where it is rapidly expanding.

Developing and emerging country demand is often met by constructing new coal, gas or oil power stations. This is because they are cheap (especially coal), the technology is relatively simple and they can be constructed quickly. Renewable alternatives have disadvantages in comparison ✓. Wind and solar power are more expensive and intermittent. HEP has long construction times and frequently involves the displacement of people to create reservoirs. Nuclear power is technically very difficult, and at up to $10 billion for one power station has very high initial costs. Cost is usually the key variable, and this makes it likely that coal and gas will be the most used fuels to meet demand in the future ✓. BP expects coal demand to increase by about 30% to 2035. China does show what is possible. In 2016 it was the world's largest user of wind power. Capacity increased from 1250 MW in 2005 to 150,000 MW in 2016, but wind still accounted for only 4% of China's total electricity generation ✓.

It is very questionable whether renewable energy will easily replace crude oil used to make transport fuels (petrol, diesel, bunker oil) that accounts for 25% of global energy use ✓. Renewable alternatives are not well developed. Biofuels can replace petrol and diesel, but they require large areas of land to grow crops, which are increasingly

20/20 marks awarded. This is a strong, well-supported answer. From the start, it shows understanding of 'future global energy demand', recognising that future demand will not be the same everywhere. This is turned into a structure for the answer by posing two sub-questions about the developed versus the developing/emerging world. This shows that the student recognises that future demand is complex..

There is a detailed discussion of wind power, its pros and cons, and the extent to which it has grown in some countries. This section is evaluative because it considers the limitations of wind power and other renewables, especially in terms of their geographic limitations and economic cost. Cost is identified as the most significant variable in the choice of which energy sources to use, which involves making a judgement, i.e. evaluating. The detail on China is a useful example that adds depth to the answer.

The section on transport fuels makes a strong argument against a renewable future. This type of clear argument is much better than sitting on the fence and shows the student is prepared to 'take the question on'.

needed to feed a growing world population. Electric vehicles are mostly charged by power stations burning fossil fuels. They could be powered by renewably generated electricity but that is a long way off in most countries.

In conclusion ✓, in developed countries renewable energy will increase its share of energy use, but total demand is static because of improving efficiency. In the developing and emerging world renewables will not meet future demand as long as fossil fuels are a cheaper option because it is easier to keep using those. There will be exceptions in some countries with the right physical conditions for wind, solar or HEP. Transport fuel demand is likely to be met by crude oil supply for the foreseeable future because no viable alternative currently exists that is widespread, although electric vehicles show promise.

> There is a clear conclusion, which returns to the theme from the introduction that future demand in the developed and developing world needs to be considered differently, as well as considering how easily a switch to renewables could be made. The answer is evaluative throughout, makes supported judgements and has a clear conclusion.

Question 10

Evaluate the view that transboundary water sources always lead to conflict between different players.

(20 marks)

> This essay question is phrased in a common way, but within it there is a 'trap'. This is that it is very easy to argue 'yes, they always lead to conflict'. After all, there are many case studies (River Nile, Aral Sea, River Ganges, Colorado River) where water is not easily shared and conflict has resulted. On the other hand, there are numerous examples of water-sharing agreements. These are equally relevant to the question and should be used to present the other side of the argument, i.e. the other 'view', before coming to an overall conclusion. Remember that 'conflict' means disagreement as well as 'war'. Most of the conflicts over water supply are wars of words, not actual war. Again, detailed place information is important to write a supported argument and conclusion (see the mark scheme levels grid for Question 5).

> **20/20 marks awarded.** This is a well-supported answer that applies examples and case studies effectively to answer the question. Beginning with an extended definition of 'transboundary' is a good way to start, as it focuses the answer on the key topic of the question. Factual detail from the UN adds weight to this.

Student answer

Transboundary water sources are those that are shared across a political boundary ✓. This includes river drainage basins, underground aquifers and lakes. Many transboundary sources straddle an international boundary such as the River Nile, but they include rivers such as the Colorado in the USA that crosses the political boundaries of US states. The UN has identified 276 transboundary international rivers ✓ and over 200 aquifers.

In most cases, transboundary water supplies do not lead to conflict ✓. According to the UN, over 450 water-sharing agreements have been signed in the last 200 years. Since 1966 ✓ the 'Helsinki Rules on the Uses of the Waters of International Rivers' have provided a legal framework to help countries resolve disputes and share water equitably. These rules were updated in 2004 to the 'Berlin Rules'.

> The answer begins by setting out its argument, that transboundary water situations are more often managed than lead to conflict. This helps avoid the answer becoming a list of 'my next example of conflict is'. The section on international rules helps justify the stated argument. Defining conflict is also sensible, as it shows understanding of the concept, i.e. it is much more than simply 'war'.

Where conflict does exist, it ranges on a spectrum from mild, diplomatic disagreement to the very rare situation when water becomes a source of open conflict ✓. In most cases, conflicts sour relations between countries. Where conflicts do exist, almost always they occur in places of existing water stress, and where other non-water-related political factors exist.

Long-standing conflict exists between India and Bangladesh over the River Ganges ✓. Low river flows in downstream Bangladesh are blamed on deforestation in the Indian Himalayas. High water pollution levels result from India using the Ganges as a human and industrial sewer. The construction of the Farakka Barrage in 1972 allowed India to divert 10% of the Ganges' flow towards Calcutta, causing reduced water availability in Bangladesh. Broader political relations between Hindu India and Muslim Bangladesh have never been good, so conflict over the Ganges has to be seen in this context ✓. A new 30-year agreement was reached in 1996, showing that some progress on sharing could be made.

> The case study on the River Ganges has the right level of detail so it avoids being descriptive. Towards the end of this section there is an evaluation that recognises the existence of an agreement — which supports the idea that conflicts can be resolved.

In Egypt and Sudan, water is a precious resource. Cairo in Egypt receives only 25 mm of rainfall per year and the country depends almost entirely on the River Nile for its water supply ✓. Historic agreements from the colonial era gave Egypt and Sudan rights to all of the Nile's waters. Today, these do not reflect the reality of population or development in upstream countries such as Ethiopia and Uganda. Upstream countries signed the Nile basin Initiative in 1999 but Egypt and Sudan refused to be involved. Increased water usage upstream, and HEP dams in Ethiopia, risk reducing Nile River flows reaching Egypt and Sudan. This is increasingly a source of tensions and, long term, could lead to open conflict, especially if Egypt feels its only water supply is threatened. A similar situation exists on the Mekong River in Asia ✓. Upstream dam construction by China risks the water supply to downstream Vietnam, Laos and Cambodia. The latter countries are part of a water-sharing treaty called the Mekong River Commission, but China is not. In both the Mekong and Nile cases the long-term solution is for all transboundary basin countries to enter into an agreement based on the existing Berlin Rules.

> The level of detail provided on the River Nile is about right, and the use of the Mekong example is good as it supports the argument being made about the Nile, i.e. that conflict exists when all parties are not part of an agreement. The River Jordan example makes the useful point that physical conflict is very much the exception not the norm.

Transboundary water supplies have contributed to armed conflict only once ✓. Between 1964 and 1967 there was a series of military clashes between Israel and its Arab neighbours (Syria, Palestine) over control of the River Jordan. Even this conflict has had some resolution and Israel and Jordan signed a water-sharing agreement in 1994.

> The final judgement, that transboundary water conflicts exist when there are water shortages and/or pre-existing unrelated political issues, recognises the complexity of the situation.

In conclusion ✓, it is not the case that transboundary water supplies always lead to conflict. There are more examples of water-sharing agreements than conflicts, and recognised international frameworks for resolving disputes. Where conflict does exist it exists as part of wider political disputes, and almost always in places with limited water supplies and no alternative supply.

Knowledge check answers

1 Oceanic plates are more dense than continental plates, and oceanic plates are thinner than continental plates.

2 Scientists have no direct observations of tectonic processes occurring inside the Earth, so it remains a theory.

3 An earthquake is a release of stored energy, which travels through the Earth as a series of waves.

4 Convergent boundaries/subduction zones produce explosive eruptions with multiple hazards.

5 The sea bed has to be displaced vertically, either up or down.

6 A natural disaster.

7 They are logarithmic scales (non-linear).

8 Subduction zones (destructive plate margins) can generate earthquakes of magnitudes greater than 9.0.

9 Landslides are very common, especially in mountainous areas.

10 Owing to the higher population density in urban areas, meaning more vulnerable people.

11 Numbers affected.

12 High magnitude, rare events, impacts on more than one country.

13 No. The level of risk can be forecast, but precise times and locations (a prediction) are not possible.

14 The recovery stage.

15 Modify the loss, because loss occurs only after the disaster.

16 The orbital eccentricity cycle, lasting 100,000 years, has the largest impact.

17 The Little Ice Age cold phase is usually linked to the Maunder Minimum.

18 The Devensian glacial period.

19 Warm-based glaciers have water at their base.

20 The Antarctic ice sheet is the largest single ice mass on Earth.

21 Freeze–thaw, or frost-shattering, weathering.

22 Firn is the intermediate stage in ice formation, between freshly fallen snow (névé) and ice.

23 Accumulation.

24 The increase in global average air temperature caused by global warming.

25 Basal slip (sometimes called basal sliding).

26 Approximately 12,500 years ago.

27 Chattermarks are produced by crushing, striations are produced by abrasion.

28 A pyramidal peak.

29 The steep (crag) sides point in a different direction: upstream for crag-and-tail and downstream for roche moutonnée.

30 A rock or boulder moved by ice from its area of origin to a new location.

31 Fluvioglacial because flowing water aligns sediment particles in the direction of flow.

32 Mountains, ridges, passes and lakes.

33 Lichens, mosses and springtails.

34 A glacial outburst flood, most commonly caused by volcanic eruption under an ice mass.

35 Glacial meltwater is an important source of water supply in many parts of the world.

36 It is a worldwide risk, so everywhere is potentially affected by it.

37 A secondary coast.

38 The layers are technically termed strata.

39 Caves are often formed because of weaknesses in the cliff.

40 Sandstone is a clastic rock, formed from cemented sand particles.

41 Plant roots bind sediment together and leaves protect the sediment surface from erosion.

42 Waves begin to break when water depth is roughly half the wavelength.

43 Summer beach profiles are steeper as they have a depositional berm at the back.

44 Longshore drift.

45 The sediment cell consists of inputs (sources), processes (transfers) and outputs (sinks).

46 Carbonation, hydrolysis, oxidation.

47 Eustatic; isostatic changes are localised.

48 A ria, or a fjord.

49 The IPCC suggest between 28 cm and 98 cm by 2100; many scientists consider 1 metre likely.

50 During major storms.

51 Storm surges are caused by low air pressure.

52 Coastal flooding.

53 Groynes because their purpose is to trap sediment moving along the coast.

54 A Shoreline Management Plan (SMP).

55 No active intervention.

56 To determine whether there is an economic case for the defences, i.e. the benefits outweigh the costs.

57 It has no external inputs or outputs.

58 Changes in mean annual precipitation or temperature.

59 Rivers and lakes (natural and artificial).

60 It affects temperatures and therefore rates of evaporation and transpiration.

61 Throughflow is through the soil; groundwater flow occurs below the water table.

62 Less runoff; more infiltration, throughflow, evaporation and transpiration.

63 Precipitation likely to exceed potential evapotranspiration throughout the year.

64 The Yenisei has a single, accentuated peak, produced by summer snowmelt and summer rain. The Rhône shows a much less accentuated summer peak produced by snowmelt in summer and winter rainfall.

65 Anything that delays or reduces runoff — vegetation cover, depth of soil, gentle slopes.

66 Local residents, local government officials (planners, engineers, etc.), developers, insurance companies.

67 (a) India and Southeast Asia; (b) west coast of South America.

68 Population growth because this increases the other three factors.

69 Intense orographic rainfall; steep slopes and fast runoff.

70 Through tsunami; vulcanicity beneath glaciers and ice sheets; blocking of rivers by lava flows and landslides induced by earthquakes.

71 Hard engineering works against nature, soft engineering with nature.

72 Important factor in determining vegetation cover and agricultural productivity.

73 Oceans.

74 India, Pakistan, Sudan, Somalia, South Africa.

75 The volume of accessible water resources in a country or region.

76 Upstream is towards the river source, downstream is towards the mouth.

77 Water needed to sustain higher living standards — more water for washing, cleaning and cooking; more recreational use of water (gardens, swimming pools).

78 Jordan: Syria, Lebanon, Jordan and Israel; Tigris–Euphrates: Turkey, Syria and Iraq; Indus: India and Pakistan; Ganges: Bangladesh and India.

79 High capital costs and technological input; not every country has a coast.

80 Store: atmosphere (sedimentary rocks, soil, etc.); flux: photosynthesis (transpiration, burning, etc.).

81 Biological, physical and carbonate.

82 Climate, vegetation cover, soil type and land use.

83 Carbon dioxide, methane and nitrous oxide.

84 Photosynthesis.

85 Coal, lignite, crude oil and gas.

86 Plants and animals are adapted to survive in particular environments. Climate is a major environmental factor. So when climate changes, so do the related flora and fauna.

87 Energy resources that are capable of being exploited, given the level of demand and available technology.

88 Irrigation, horticulture and intensive livestock rearing.

89 Forms of waste that can be re-used to produce more energy — nuclear fuel, heat from manufacturing processes, etc.

90 It is clean and easily moved by transmission cables.

91 Much modern technology is energy-consuming. Rising standards of living usually result in more energy-consuming appliances in the home.

92 Because the USA has much larger accessible reserves of fossil fuels than France.

93 More accessible fields are becoming exhausted; price of oil does not warrant exploitation of less profitable fields.

94 Usually on the grounds of unsightliness; noise in the case of wind farms.

95 Deindustrialisation.

96 A carbon sink has the ability to absorb carbon dioxide; a carbon store has the ability to retain carbon for varying periods of time.

97 Taiga forest is more extensive and only exploited for its timber.

98 A point in time when change becomes irreversible and moves from one stable state to another.

99 The short-wave solar radiation that is reflected from the Earth back into space.

100 Land-use planning examples: soft-engineering — restricting development in vulnerable areas; hard engineering — building structures to make areas less vulnerable.

101 Improving energy efficiency or carbon taxation.

102 Climate change is a global problem; no one country can mitigate it. It needs co-ordinated international action.

Note: page numbers in **bold** indicate location of key term definitions. Some of the page numbers will refer to illustrations.